I0469976

EXTERIOR BALLISTICS

EXTERIOR BALLISTICS

A New Approach

George Klimi

Copyright © 2010 by George Klimi.

Library of Congress Control Number: 2010903249
ISBN: Hardcover   978-1-4500-5910-7
                     Softcover   978-1-4500-5909-1
                     Ebook   978-1-4500-5911-4

All rights reserved. No part of this book may be reproduced or transmitted in any form or by any means, electronic or mechanical, including photocopying, recording, or by any information storage and retrieval system, without permission in writing from the copyright owner.

This book was printed in the United States of America.

In the front cover is displayed a collage of the photos of sculptures of the great sculptor J. Paco (1914-1991).

To order additional copies of this book, contact:
Xlibris Corporation
1-888-795-4274
www.Xlibris.com
Orders@Xlibris.com
72305

To the good friends of mine

To my beloved hometown, TIRANA

To the wonderful NEW YORK CITY

Contents

ACKNOWLEDGMENTS

I would like to thank my colleagues at the math department in City Tech and the NYC College of Technology for the support to my research and educational activity.

Special thanks to Jan Krčmář, PhD (physicist, Czech Republic) for his kind permission to download and use his excellent PC program Ballistica2.2 as well as for his comments on some problems of exterior ballistics we have discussed together.

I must acknowledge with great appreciation the encouragement and support of my kind-hearted friend and colleague Jack Lowenthal, PhD.

INTRODUCTION

T he book *Exterior Ballistics: A New Approach* presents the exterior ballistics of flight of point-mass projectiles introducing an innovative approach based on the analytical G-drag functions (G_1, . . . G_8, etc.) and projectile trajectory-streamline and Snell's law model.

The analytical G-functions of resistance give to ballisticians analytical formulae for the trajectory of the projectile flight and the possibility to integrate in quadratures the systems of differential equations that describe the projectile flight.

Employing analytical G-functions, we have created simple universal PC programs that enable the reader to easily solve a variety of theoretical and practical ballistics problems related with the point-mass projectile flight in standard or nonstandard atmosphere and in presence of wind.

Among other methods in the study of flight of projectiles, *Exterior Ballistics: A New Approach* introduces as well the method based on the "projectile trajectory-streamline and Snell's law model" that is the result of the assumption that a projectile demonstrates as well waves properties.

The projectile trajectory-streamline and Snell's law model is a fundamental result of my recent findings in exterior ballistics, obtained by applying Sir Isaac Newton's postulate on the wave nature of moving bodies (point-mass particles) and his interpretation of Snell's law on refraction of waves to the flight of projectiles.

The projectile trajectory-streamline and Snell's law model is based on two assumptions:

- The projectile trajectory is assumed to be a streamline of point-mass projectiles (bullets) moving through the superimposed gravitational field and air field of flow.
- The point-mass projectile, passing through the "interface" that separates two air environments, with different acoustics impedances, demonstrates as well wave properties and follows the path shown by Snell's law of refraction.

The projectile trajectory-streamline and Snell's law model enables us to find the elements of the projectile trajectory with simple mathematical formulae without solving the differential equations that describe the ballistics trajectory of the projectile.

The projectile trajectory-streamline and Snell's law approach can be applied to mountain or high-altitude shootings, to winter low-temperature shootings, or to summer high-temperature shootings as well as to construct nonstandard firing range tables and find the range corrections.

The results obtained solving exterior ballistics problems using the projectile trajectory-streamline and Snell's law model are confirmed employing other solution methods or firing range tables.

The impressive outcomes we obtain solving the exterior ballistics problems employing the fictive projectile trajectory-streamline and Snell's law model demonstrate at the same time the verity of Newton's postulate on the dual nature of macroscopic bodies, i.e., that the body in flight manifests as well the properties that can be explained using the wave model of a moving particle.

The projectile trajectory-streamline and Snell's law model is an original contribution that broaden the modern methods of studies in exterior ballistics and might open new perspectives in the study of aerodynamics.

Exterior Ballistics: A New Approach is addressed to professional ballisticians, military faculty and students, researchers in the field of ballistics and gunnery, as well as physics, mathematics, forensic and computer sciences faculty and students.

A good level of knowledge in undergraduate general physics and calculus and some knowledge in differential equations is needed to understand the material and apply the new methods in the solution of exterior ballistics problems.

In the first chapter of the book, we present the foundations on which the *Exterior Ballistics: A New Approach* is developed in the following three chapters. We introduce as well the analytical G-functions of resistance and show the way they are obtained.

The analytical G-drag functions allow us to solve throughout the book a variety of exterior ballistics problems relatively easy, numerically or in analytical way, avoiding the traditional use of tabulated data of different tabular G-drag functions that are employed nowadays in exterior ballistics.

The second chapter contains different solution methods of the differential equations that describe the projectile flight to the target. Here we introduce the projectile trajectory wave and Snell's law model, demonstrating at the same time truthfulness and accurateness of the original model in the solution of the exterior ballistics problems.

There are presented as well four universal PC programs that are used to solve numerically the differential equations of projectile flight of unguided projectiles in standard or nonstandard atmosphere and solve the main problems of exterior ballistics of unguided projectiles.

The third chapter includes some approximate methods of solutions of the differential equations of projectile flight, among them the Siacci's method that is applied to any G-function of resistance allowing us to obtain analytical formulae.

In the fourth chapter are presented some methods of the theory of corrections, particularly the correction method based on the use of the projectile trajectory-streamline and Snell's law model that allow us to determine range and departure angle corrections.

The book is organized in a methodical way, not only to convey the new and the updated information in a simple theoretical way, but also to illustrate the new approach to exterior ballistics with examples from the practice of firing with artillery and small arms.

The PC programs are as well demonstrated through the solution examples and instructions to make them easy to be used.

The solutions of exterior ballistics problems that are obtained using any of the theoretical methods presented in the book are compared most of the time with the solutions of the same problems obtained by other authors or the solutions that are contained in the range tables of the firearms.

At the end of each section (where it is possible) there are formulated and solved examples associated with the respective solutions. There are around eighty solved examples that illustrate the theoretical methods of solutions of exterior ballistics problems.

Exterior Ballistics: A New Approach is a unique book in the literature of exterior ballistics for the original methods introduced to solve the exterior ballistics problems and particularly for the projectile trajectory-streamline and Snell's law model.

The present book, together with my other two books, *Exterior Ballistics with Applications* and *Exterior Ballistics of Small Arms* (already published by Xlibris) make a comprehensive reference material to the study of exterior ballistics and to find solutions to the ballistics problems encountered in theory and practice of shooting.

Because the exterior ballistics literature during the cold war was somehow considered classified, the modern exterior ballistics literature that is published and made available to the general public nowadays is very limited.

Great reference sources of information for writing the three books on exterior ballistics have been the authoritative books:

- *Modern Exterior Ballistics* by Robert L. McCoy, Schiffer publishing, 1999

- *Exterior Ballistics* by J. M. Shapiro, Oborongiz, 1950
- *Foundations of Ballistics* by B. N. Okunev, Vol. 1, National Commission of Defense, 1943
- *Handbook of Ballistics* by C. Cranz, London, 1921

As a summary, *Exterior Ballistics: A New Approach* is, contains or introduces

- new methods and PC programs to solve practically all the main problems of exterior ballistics of unguided projectiles;
- an excellent reference material that provides answers to problems encountered in the practice of motion of unguided projectiles fired by artillery and small arms;
- four compact types of original universal PC programs that enable the ballisticians to solve any problem of the exterior ballistics of unguided projectiles and the problems of the fire control;
- original analytical drag functions G_1, G_2, G_7, G_8, Ingalls', etc., that lay the foundation of *Exterior Ballistics: A New Approach*;
- the original projectile trajectory-streamline and Snell's law model in the study of exterior ballistics;
- a relative large number of illustrative solved exterior ballistics problems.

The Wonders of Snell's Law

Doing research on Exterior Ballistics, for the projectile trajectory in nonstandard atmosphere I obtained some results that could not be explained using Huygens' Principle of Snell's law

$$\frac{\sin \alpha_1}{\sin \alpha_2} = \frac{v_1}{v_2},$$ (1)

but Newton's interpretation of Snell's law, [1]

---

1. Prof. Walter H. G. Lewin, Snell's Law, http://videolectures.net/mit802s02_ lewin_lec29/ (web access on June -August, 2009)

$$\frac{\sin\alpha_1}{\sin\alpha_2} = \frac{v_2}{v_1}, \tag{2}$$

that I was not aware of its existence.

Looking for Snell's law, I found and "attended" Prof. Walter Levin's video-lecture on Snell's law (lecture 29th). I was able to lay the foundations for my assumptions on the flight of projectiles in air, and develop a new model and an original approach to solve the Exterior Ballistics problems with simple mathematics formulae.

I went further interpreting Newton's statement on Snell's law (2) considering the projectile as a streamline of point-mass particles launched with a given departure speed and departure angle.

The model of the "projectile trajectory-streamline and Snell's law" gives remarkably correct results solving the two main problems of Exterior Ballistics:

- find the departure angle for a given range;
- find the range for a given departure angle.

I believe that the impressive outcomes we obtain solving the exterior ballistics problems employing the projectile streamline-wave and Snell's law model, demonstrate at the same time the verity of Newton's postulate on the dual nature of macroscopic bodies, i.e. that the bodies in flight manifests as well the properties that can be explained using particle behavior of waves.

I would better say that Snell's law (2), as it is interpreted by Newton, is valid for particles and it is in conformity with the laws of conservations of mass, momentum and energy though it is accepted that Snell's law is a property of waves, and particularly of light (see section 2.7).

Gjergj Klimi, PhD.
New York, January 21, 2010

ABOUT THE AUTHOR

George Klimi (Gjergj Klimi) earned his BA and PhD degrees in Physics from the University of Tirana. At present, he is assistant professor of mathematics at the NYC College of Technology and adjunct faculty at Pace University (New York).

George is the author of the books *Exterior Ballistics with Applications* (2008) and *Exterior Ballistics of Small Arms* (2009), published by Xlibris.

From 1970 till 1993, George Klimi has been professor of physics and mathematics at the Military Academy of Tirana. In the '80s he was chair of the physics department and at the same time adjunct faculty at the Military Academy of General Staff in Tirana.

He taught physics, calculus, probability applied in antiaircraft firing and lectured on the applications of physics in conventional and nuclear explosions.

In the '90s, George used to work at the Committee of Science and Technology, and at the Ministry of Higher Education and Research in Albania as Director of Tempus Office, responsible for the implementation of the European Community Programs to restructure the higher education in Albania.

He was awarded the Gold Medal of Eagle by the President of Albania for his contribution in the democratic movement against the communist dictatorship and for restructuring the higher education.

ELECTRONIC COPIES OF PC PROGRAMS

Dear Reader,

To request an electronic copy of the PC programs that are presented in the book, please send a message to iven24@aol.com, gklimi@pace.edu, or at gklimi@citytech.cuny.edu

DISCLAIMER

Neither the author nor Xlibris accept any responsibility for any problem (errors, damages, injuries, etc) that might occur in practice of shooting applying the PC programs or other methods, presented in the book, to solve practical exterior ballistics problems.

Though the PC programs most of the time give good outcomes, it might occur that the reader obtains wrong solutions, for example as result of incorrect input of data, calculation of erroneous preliminary data needed to run a program, incorrect use of PC programs, use of the data given by different authors, use of data obtained experimentally, etc.

The user of the PC programs and other methods of solutions of exterior ballistics problems should be cautious and use the common sense to judge the results obtained using the PC programs or the other solution methods.

1

THE EQUATIONS OF PROJECTILE FLIGHT

Introduction

In the present chapter, we lay the foundations on which the *"Exterior Ballistics: A New Approach"* is developed introducing a set of analytical G-functions of resistance.

We use analytical G-drag functions to solve the differential equations of projectile flight and a large number of exterior ballistics problems relatively easy, avoiding the traditional use of tabulated data of different tabular G-drag functions that are employed nowadays in exterior ballistics.

1.1. Drag Force and Drag Acceleration

We study the projectile flight with respect to a three-dimensional system of rectangular Cartesian coordinates that has the origin at the sea level.

We assume that the projectile is launched from the muzzle of the firearm (located at the origin of the coordinates) with an initial velocity $\vec{v}_0$ that makes an angle α_0 with horizontal line (figure 1). The y-axis is at the same plane with the x-axis and the projectile initial velocity $\vec{v}_0$. The z-axis is perpendicular to the launching plane.

The target T is on the x-axis at the horizontal range x_T.

Geometric Characteristics. Elements of Ballistics Trajectories

The projectile trajectory (ballistics trajectory) is the path that the projectile follows from the muzzle to the target.

In general, considering the ballistics trajectory in the x-, y- coordinative plane (there is no deviation of the trajectory along z-axis; for example there is no cross-wind), the geometric elements related to the ballistics trajectory with respect to a rectangular Cartesian system of coordinates are presented in figure 1, figure 2, and figure 3. The elements of the projectile trajectory are:

- the horizontal line (plane) is the line (plane) parallel to the x-axis at the muzzle of the firearm;
- the departure line is the tangent line to the ballistics trajectory at the muzzle and has the direction of the initial velocity $\vec{v}_0$ of the projectile. (the initial speed of projectile is v_0);
- the angle of departure α_0 is the angle measured from the horizontal line (plane) to the line of departure;
- the line of sight (LOS) is the straight line that connects the muzzle of firearm and the target (center or a point on the target);
- the angle of sight (A_S) is the angle measured from the horizontal plane to the line of sight. (It is called also elevation when the target is at a level above the level of the muzzle, and depression otherwise);
- the angle of projection A is the angle measured from the line of sight to the departure line;.
- the line of elevation is the line "oj" that coincides with the axis of the bore;
- jump is the angle measured from the line of elevation to the departure angle;
- quadrant elevation is called the angle measured from the horizontal line to the elevation line.

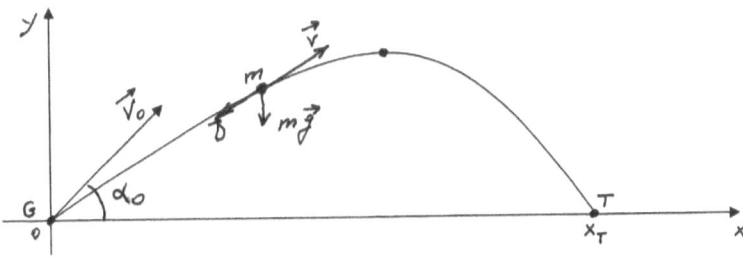

Figure 1

We assume that the projectile is launched from the muzzle with an initial velocity $\vec{v}_0$ that makes an angle α_0 with horizontal line (figure 1 and figure 2), or that the projectile is lunched at the muzzle located over the origin of coordinates at point G with ordinate y_0, figure 3.

The coordinates of the point of impact, called also the coordinates of the target T, are (x_T, y_T). In figure 1, the target T is on the x-axis at the horizontal range x_T. In figure 2 and in figure 3, the target is at the point with coordinates (x_T, y_T).

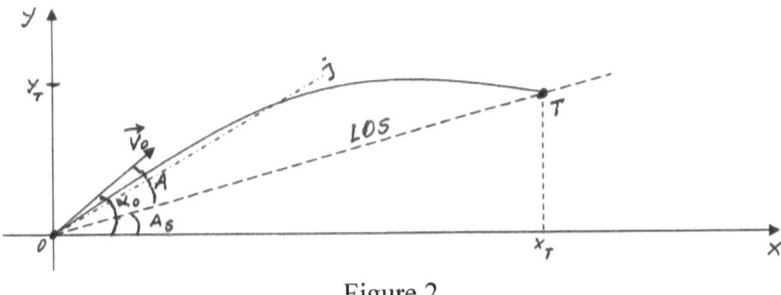

Figure 2

The *horizontal range* x_T is the distance from the muzzle of the firearm to the target T, when both target and the muzzle of firearm are at the sea level or at the same altitude.

In figure 2 and in figure 3, the distance from the firearm to the target (OT, or GT) is called the inclined range.

The terminal speed v_T is the speed of the projectile at the target.

The angle of impact on the target is α_T, the time of flight to the target t_T, and the maximum trajectory altitude is y_{max} (coordinates of the trajectory vertex are (x_m, y_m)).

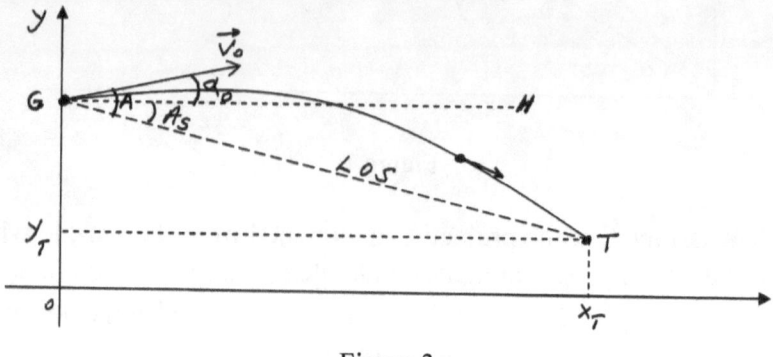

Figure 3

The Main Problems of Exterior Ballistics

Using the methods of exterior ballistics, we can determine all the elements of a trajectory at a given point if there are known three of them mainly three out of four $(\alpha_0, v_0, c, (x_T, y_T))$ trajectory characteristics.

The main problems the exterior ballistics deals with are the following:

- For a given ballistics coefficient c and a given departure velocity $\vec{v}_0$ (initial speed v_0, departure angle α_0) of a projectile, find the range of fire x_T (or the coordinates of the point of impact (x_T, y_T) to the target), and all the other elements of the trajectory at the impact point;

- For a given departure speed v_0 of a projectile, a given range of fire x_T (or a given point on the trajectory), and a given ballistics coefficient c, find the departure angle α_0 and all the other trajectory elements of the trajectory at the impact point;

- For a given departure angle α_0, given initial speed v_0, given range x_T (given point of impact (x_T, y_T)), find the ballistics coefficient c and all the other elements of the projectile trajectory.

- For a given departure angle α_0, range x_T, and ballistics coefficient c, find the initial speed v_0, and the rest of the elements of the projectile trajectory.

The Standard Atmosphere

The study of the projectile flight in the real atmosphere is complicated by the variations of the density, pressure, and temperature of air with altitude and time, as well as by the presence of water vapor, wind, rain, snow, etc.

To simplify the study of the projectile trajectory, and the solution of the differential equations of projectile flight, the exterior ballistics considers some standard characteristics for the atmospheric air, where the density, the pressure, and the temperature of air depend only on the altitude over the sea level.
The wind velocity is absent.

Hereafter, there are defined three standard atmospheres, the ASM atmosphere, the ICAO atmosphere, and TSA atmosphere that are in use in exterior ballistics.

Army Standard Metro (ASM) Atmosphere

The ASM atmosphere have the following characteristics [2]:

- at the sea level ($y=0$) the air density is $\rho_{0N}=1.2034 kg/m^3$, the temperature of air is $T_{0N} = 15\ ^{\circ}Celsius$, the atmospheric pressure is $p_{0N}=750$ mm Hg., the speed of sound $a_{0N}=341.458 m/s$, and corresponds to 78% of the relative humidity and to the virtual temperature approximately $\tau_{0N}=289.60 K$;[3]

---

2. In "Exterior Ballistics with Applications" the Standard Atmosphere, is somewhat different from the Army Standard Metro (ASM); we call it TSA atmosphere.
3. McCoy, Robert, "Modern Exterior Ballistics", page 101 and page 169, Schiffer Publishing, 1999.

- the temperature t, the pressure p, and the density ρ of the atmospheric air changes (decreases) with the altitude y above the sea level;
- the projectile flight is in absence of wind.

ICAO Standard Atmosphere

The International Civil Aviation Organization atmosphere, (ICAO atmosphere) [4] has the following characteristics:

- at the sea level ($y=0$), the air density is $\rho_{0N}=1.2251 kg/m^3$, the temperature of air is $T_{0N} = 15\ °Celsius$, the atmospheric pressure is $p_{0N}=760mmHg$., the speed of sound $a_{0N}=340.30m/s$ corresponds to the relative humidity 0% and to the virtual temperature equal to the temperature of dry air, $T_{0N}=288.15K$;
- the temperature t, the pressure p, and the density ρ of the atmospheric air changes (decreases) with the altitude y above the sea level;
- the projectile flight is in absence of wind.

Traditional Standard Atmosphere (TSA)

In *Exterior Ballistics with Applications*", we assume that the projectile motion is in a standard atmosphere, with the following characteristics:

- at the sea level ($y=0$), the air density is $\rho_{0N}=1.205kg/m^3$, the temperature of air (virtual temperature) is $\tau_{0N}=289.08K$ and corresponds to the relative humidity 50%, the atmospheric pressure is $p_{0N}=750$ mm Hg., the speed of sound is $a_{0N}=340.83m/s$;

---

4.   In "Exterior Ballistics with Applications" the Standard Atmosphere, is somewhat different from the Army Standard Metro (ASM), and ICAO standard Atmosphere.

- the temperature, the pressure, and the density of air vary only with the altitude "y" above the sea level. The variation of the temperature with altitude is linear;
- the projectile flight is in absence of wind.

Notes on the Standard Atmosphere

In exterior ballistics, there are differences between the characteristics of "standard atmospheres" that are in use.

- The ASM atmosphere considers a standard pressure of $p_{0N} = 750mmHg$. 78% relative humidity, speed of sound $a_{0N} = 341.458 m/s$, temperature $t_{0N} = 15°C$, and air density $\rho_{0N} = 1.2035 kg/m^3$.
- The ICAO atmosphere considers a standard pressure of $p_{0N} = 760mm\ Hg.$, relative humidity 0%, speed of sound $a_{0N} = 340.30 m/s$, temperature $t_{0N} = 15°C$, and air density $\rho_{0N} = 1.225 kg/m^3$.
- Mayevski's standard atmosphere (according to Alger, P. R, *"The Groundwork of Practical Naval Gunnery"*, p.37, p.39, 2nd edition, 1917) assumes $p_{0N} = 750mm\ Hg.$, $t_{0N} = 15°C$, and relative humidity 50%.
- In *"Exterior Ballistics with Applications"*, we consider the TSA atmosphere $p_{0N} = 750mm\ Hg.$, $a_{0N} = 340.83 m/s$, $t_{0N} = 15°C$ and a relative humidity of 50%.
- Gubin, S. G, and Gorovoj, S. A in *"Ballistika"*, chapter 4, section "Atmospheric Pressure" (http://www.pishtov.com/ballistic/pres.htm), considers as normal the pressure $p_{0N} = 750mm\ Hg.$ at the firing site 110 meters over the sea level. In this case, the corresponding sea level atmospheric pressure is approximately $p_0 = 760mm\ Hg$. It is assumed that the departure point of the projectile is on the ground at the altitude 110 meters, and not at the sea level. However, in some Russian artillery range tables, at the sea level, it is assumed an atmospheric pressure $p_{0N} = 750mm\ Hg$.

- Ingalls in his *"Explanatory Notes"*, Ingalls' Ballistics Tables (1909), considers a standard atmosphere with pressure $p_{0N}=750mm\ Hg.$, temperature $t_{0N}=15°C$, relative humidity 50%, i.e. practically the TSA.

Point-Mass Projectile

The traditional exterior ballistics assumes that the projectile is a point-mass with the following characteristics:

- The projectile is a particle with constant mass m located at the projectile center of masses.
- The ballistics characteristics of projectile: mass m, caliber d, are standard and identical to the manufacturing characteristics.
- The projectile does not rotate about its axis of symmetry. The projectile axis at any moment in flight has the direction of the projectile velocity $\vec{v}$.
- The projectile cross-sectional area A is perpendicular to the projectile velocity $\vec{v}$.
- Earth does not rotate and its surface is plane.
- The gravitational acceleration $\vec{g}$ is constant with magnitude $g=9.80665m/s^2$.
- The forces acting on the projectile at any moment in flight are: the gravitational force $\vec{F}_G=m\vec{g}$ and the resistance of air (drag force) $\vec{D}$ opposite to the projectile velocity.

The projectile trajectory, that is described based on the above assumptions of the point-mass projectile, is called the "Point-Mass Trajectory".

The Standard Temperature of Propellant Charge

For the ICAO atmosphere and for the ASM atmosphere, we should note that the standard temperature of propellant charge is considered as 21 degrees Celsius. The standard temperature of the propellant charge for the TSA atmosphere is considered as 15 degrees Celsius.

Drag Force and the Drag Coefficient

The magnitude of the drag force that air exerts on a point mass projectile is

$$D = A\frac{\rho v^2}{2}C(\frac{v}{a}),\qquad(1.1.1)$$

while the drag force is

$$\vec{D} = -A\frac{\rho v^2}{2}C(\frac{v}{a})\frac{\vec{v}}{v},\qquad(1.1.2)$$

where A is the cross-sectional area of the projectile, ρ is the air density, v is the speed of the projectile, a is the speed of sound in air, and $(\vec{v}/v)$ is the unit vector in the direction of the projectile velocity $\vec{v}$. The temperature, density, and pressure change with altitude over the see level.

The quantity $C(v/a)$ is the drag coefficient that is a function of the dimensionless Mach number,

$$M = v / a.\qquad(1.1.3)$$

The drag coefficient $C(v/a)$ is a characteristic of flight of a given projectile.

In general, the drag coefficient $C(v/a)$ is different for different projectiles. The drag coefficient $C(v/a)$ can be determined experimentally for each projectile, for example using the "wind tunnels", Doppler radars, or employing other ballistics methods[5].

However, in practice, instead of determining a drag coefficient $C(v/a)$ for any projectile, the exterior ballistics finds experimentally a drag coefficient $C_D(v/a)$ for a standard projectile and uses that for a large variety of projectiles.

The drag coefficient $C_D(v/a)$ is a function of Mach number, presented in tabular form, called the "table of standard drag

---

5. Useful information the reader can find in the e-book "Fundamentals of Ballistics", by Prof. Gunther Dyckmans (2007-2009); https://e-ballistics.com/ebook

coefficients" (for example, the drag coefficient of Mayevski-Zabudski's function presented in table 1, section 1.5 below).

The drag coefficient $C(v/a)$ of a non standard projectile is related to the drag coefficient of a standard projectile, $C_D(v/a)$, by the equation:

$$C(\frac{v}{a}) = iC_D(\frac{v}{a}),$$
(1.1.4)

where i is the "form factor" or the "form coefficient" of the projectile.

The form factor i is a characteristic of the shape of the projectile. It is a variable quantity that depends on Mach number $M=(v/a)$ and on the particular standard drag coefficient $C_D(v/a)$ we use in (1.1.4) to determine $C(v/a)$.

Substituting (1.1.4) in (1.1.1), we find that the magnitude of the drag force "D" exerted on a non-standard projectile can be written through the drag coefficient $C_D(v/a)$ of the standard projectile as follows:

$$D = A\frac{\rho v^2}{2}iC_D(\frac{v}{a}).$$
(1.1.5)

Thus, the drag force that is exerted on the projectile is

$$\vec{D} = -A\frac{\rho v^2}{2}iC_D(\frac{v}{a})\frac{\vec{v}}{v},$$
(1.1.6)

1.2. Vector Differential Equation of Projectile Flight

Newton's second law for a given point-mass projectile in flight with respect to a three-dimensional Cartesian coordinate system (figure 1) can be written:

$$m\frac{d^2\vec{r}}{dt^2} = m\vec{g} - A\frac{\rho v^2}{2}iC_D(\frac{v}{a})\frac{\vec{v}}{v},$$
(1.2.1)

where $\vec{r} = \vec{r}(t)$ is the position vector of the projectile at time t, and i is the form factor.

- The trajectory of a point-mass projectile is a plan curve in the plane of flight (x-, y-plane). At any instant during the flight, the z-coordinate of the point-mass projectile, in absence of crosswind, is zero.

From equation (1.2.1), we obtain the following differential equation in vector form:

$$\frac{d\vec{v}}{dt} = \vec{g} - \frac{1}{m} A \frac{\rho v^2}{2} i C_D(\frac{v}{a}) \frac{\vec{v}}{v}. \tag{1.2.2}$$

For the standard projectile, the form coefficient is one, i.e. $i = 1$.

Drag Acceleration

The projectile acceleration that is result of the drag force (1.1.6) exerted on the projectile by the atmospheric air, is

$$\vec{a}_D = \frac{D}{m} = -A \frac{\rho v^2}{2m} i C_D(\frac{v}{a}) \frac{\vec{v}}{v}. \tag{1.2.3}$$

The magnitude of the projectile acceleration is

$$a_D = A \frac{\rho v^2}{2m} i C_D(\frac{v}{a}). \tag{1.2.4}$$

It is important to recall that the drag force does not depend on the fact that the projectile is moving in motionless air with velocity $\vec{v}$, or that the projectile is at rest while the air is moving in opposite direction with velocity $(-\vec{v})$. In other words, the magnitude of the drag force is a function of the relative speed of the projectile with respect to air.

The above property of the drag force is used in the wind tunnels to measure the drag coefficient of a projectile.

Substituting the cross-sectional area of the projectile, $A = \pi \cdot d^2 / 4$ (where d is the diameter of the cross-sectional area A of the projectile), in (1.2.4) we obtain:

$$a_D = \frac{\pi}{8} \cdot \frac{id^2}{m} \cdot \rho v^2 \cdot C_D(\frac{v}{a}),$$

or

$$a_D = 0.3927 \cdot i\frac{d^2}{m} \cdot \rho v^2 \cdot C_D(\frac{v}{a}) \qquad (1.2.5)$$

Multiplying and dividing the right side of the above equation by $1000\rho_{0N}$, where ρ_{0N} is the density of air at the sea level, we can write the drag acceleration as follows:

$$a_D = (3.927 \times 10^{-4} \cdot \rho_{0N})\frac{id^2}{m}1000\frac{\rho}{\rho_{0N}}v^2 C_D(\frac{v}{a}). \qquad (1.2.6)$$

Denoting

$$h(y) = \rho / \rho_{0N}, \text{ and } c = \frac{id^2}{m}1000, \qquad (1.2.7)$$

for the drag acceleration (1.2.6) we have:

$$a_D = (3.927 \times 10^{-4} \cdot \rho_{0N}) \cdot c \cdot h(y) \cdot v^2 C_D(\frac{v}{a}). \qquad (1.2.8)$$

G-Drag Function

We denote

$$G_D(v) = (3.927 \times 10^{-4} \rho_{0N})v^2 C_D(\frac{v}{a}), \qquad (1.2.9)$$

the G-function of resistance of a standard projectile that moves in standard atmosphere, with density, pressure, temperature, and speed of sound are respectively ρ_{0N}, p_{0N}, τ_{0N} and a_{0N}.

Density Function

The virtual temperature of air changes with the altitude over the sea level and can be determined by the function:

$$\tau = \tau_{0N} - 0.006328y.$$

The quantity $h(y)$, called density function, for altitudes till around 10,000 meters over the sea level, can be determined approximately using the following formula:

$$h(y) = (\frac{\tau_{0N} - 0.006328y}{\tau_{0N}})^{4.4}, \qquad (1.2.10)$$

where τ_{0N} is the virtual temperature at the sea level.

For the ASM standard atmosphere the virtual temperature at the sea level is $\tau_{0N} = 289.60K$ Kelvin, while for the ICAO standard atmosphere the virtual temperature is $\tau_{0N} = 288.15K$.

Thus, the density function for the ASM atmosphere and the density function for the ICAO atmosphere are respectively:

$$h(y) = (\frac{289.6 - 0.006328y}{289.6})^{4.4}, \ h(y) = (\frac{288.15 - 0.006328y}{288.15})^{4.4}.$$

$$(1.2.11)$$

Ballistics Coefficient

The ballistics coefficient of a given projectile is

$$c = \frac{id^2}{m}1000,$$

where i is the form coefficient, d is caliber of the projectile, and m is the respective mass.

The ballistics coefficient in use in the U.S. Army, expressed in English units, lb/in^2 (pound per square inch), is

$$C = \frac{m}{i \cdot d^2}.$$

The relation between the above coefficients is expressed by the formula:

$$C = \frac{1.4222}{c}.$$

The Vector Differential Equation

The vector differential equation (1.2.2) can be written:

$$\frac{d\vec{v}}{dt} = \vec{g} - 3.927 \times 10^{-4}\, \rho_{0N} h(y) \cdot c \cdot v^2 C_D(v/a) \cdot \frac{\vec{v}}{v}. \quad (1.2.12)$$

The Drag Acceleration in the ASM Atmosphere

Substituting the sea level ASM density of air $\rho_{0N} = 1.2034\, kg/m^3$ in (1.2.6) and (1.2.9), for the acceleration of the projectile in the ASM atmosphere and the corresponding G-function, we obtain respectively:

$$a_D = (4.726 \times 10^{-4}) \cdot ch(y) v^2 C_D(\frac{v}{a}), \quad (1.2.13)$$

and

$$G_D(v) = 4.726 \times 10^{-4} \cdot v^2 \cdot C_D(v/a), \quad (1.2.14)$$

where

$$h(y) = (\frac{289.6 - 0.006328y}{289.6})^{4.4}.$$

As a matter of fact, the G-function of resistance $G_D(v)$ is a function of the projectile speed v and the speed of sound a.

The reason we denote the G-function of resistance $G_D(v)$ and not $G_D(v/a)$ will be clear in the forthcoming sections.

For the ASM atmosphere, considering (1.2.13), the vector differential equation (1.2.12) can be written in the form:

$$\frac{d\vec{v}}{dt} = \vec{g} - c \cdot h(y) \cdot G_D(v) \cdot \frac{\vec{v}}{v}, \qquad (1.2.15)$$

or

$$\frac{d\vec{v}}{dt} = \vec{g} - 4.726 \cdot 10^{-4} c \cdot h(y) \cdot v^2 C_D(v/a) \cdot \frac{\vec{v}}{v}. \qquad (1.2.16)$$

Drag Acceleration in the ICAO Atmosphere

The G-function of resistance for a projectile launched in ICAO standard atmosphere is slightly different from G-function (1.2.14). In ICAO standard atmosphere, the G-function is determined by the equation:

$$G_{DA}(v) = 4.811 \times 10^{-4} \cdot v^2 C_D(v/a), \qquad (1.2.17)$$

where the speed of sound and the air density at the sea level are respectively:

$$a_{0N} = 340.30 m/s, \quad \rho_{0N} = 1.2251 kg/m^3. \qquad (1.2.18)$$

The magnitude of the drag acceleration is

$$a_D = (4.811 \times 10^{-4}) \cdot c \cdot h(y) v^2 C_D(\frac{v}{a}). \qquad (1.2.19)$$

The vector differential equation of the projectile flight in ICAO atmosphere is

$$\frac{d\vec{v}}{dt} = \vec{g} - 4.811 \cdot 10^{-4} c \cdot h(y) \cdot v^2 C_D(v/a) \cdot \frac{\vec{v}}{v}, \qquad (1.2.20)$$

where

$$h(y) = (\frac{288.15 - 0.006328y}{288.15})^{4.4}. \qquad (1.2.21)$$

Example 2.1

Calculate the ballistics coefficient of the projectile of an 82mm-caliber trench mortar 82mm. Projectile mass is m = 3.153 kg. The form factor of the projectile (with respect to the Siacci's function of resistance) is $i=0.60$.

Solution

Substituting $m = 3.153kg$, caliber $d = 82mm = 0..82m$, and the form factor $i = 0.60$ we obtain

$$c = \frac{id^2}{m}1000 = \frac{(0.6)(0.082)^2}{3.153} \cdot 1000 = 1.28m^2 \, / \, kg$$

Example 2.2

Calculate the ballistics coefficient of the bullet of a Russian rifle of caliber d =7.62 mm, bullet mass m = 7.9g, form factor (with respect to the Siacci's function of resistance) is $i = 0.56$.

Solution

Substituting the caliber d = 7.62mm = 0.00762m, mass m = 7.9g = 0.0079kg, and $i=0.56$, we find that the ballistics coefficient of the given projectile is

$$c = \frac{id^2}{m}1000 = \frac{(0.56)(0.00762)^2}{(0.0079)}1000 = 4.116m^2 \, / \, kg \, .$$

1.3. G-Function of Resistance

The G-function of resistance (1.2.9) seems to have a complicated form because of the presence of the drag coefficient that can not be modeled perfectly by a simple mathematical function that would make simpler the integration of the differential equations of the projectile trajectory that result from (1.2.12).

Anyway, we will see that any G-function of resistance can be modeled by an approximate two-piecewise function, similar to the approximate Siacci's function of resistance,

$$K_D(v) = \begin{cases} 1.212 \cdot 10^{-4} v^2 & for \quad v \le 256 m/s \\ (v-240)/3 & for \quad v > 256 m/s \end{cases},$$

we use in *"Exterior Ballistics with Applications"*.

The G-function of resistance (1.2.9) is a function of Mach number. In other words, the G-function is a function of two variables: the projectile speed v and the speed of sound a.

To simplify the solution of the differential equations of the projectile flight, exterior ballistics reduces the dependence of G-function from two variables into only one, i.e., to the projectile speed v. (Recall that the projectile speed is equal to the relative speed of air flow with respect to the projectile. The relative velocity of air flow is in the opposite direction of the projectile.)
During the flight, the projectile speed v and the speed of sound a in atmospheric air change continuously.

Let's consider that the drag coefficient of a given projectile is

$$C_D = C_D(u/a) = C_D(M).$$

Let's assume that the projectile trajectory (the trajectory-streamline) is such that its speed v is determined by the equation

$$\frac{v}{a_{0N}} = \frac{u}{a} = M, \tag{1.3.1}$$

where the speed of sound, during the flight, is constant and equal to the speed of sound at the sea level a_{0N}.

From the above equations, it follows that we are assuming that the air field of flow has kinematics similitude [6].

---

6.   Any two air flows (streamlines) in a field of flow have kinematics similitude when the ratio of corresponding particles velocities is constant. For the projectile, we consider streamline the relative flow of air with respect to the projectile.

Thus, using the kinematics similitude, we can consider that the drag coefficient is obtained for a projectile that flies in a given standard atmosphere (not necessary ASM atmosphere or ICAO atmosphere) where, at the sea level, the speed of sound is a_{0N}, the virtual temperature is τ_{0N}, the pressure is p_{0N}, and the density is ρ_{0N}.

The G-function, determined in (1.2.9), is

$$G_D(v) = (3.927 \times 10^{-4} \rho_{0N}) v^2 C_D(\frac{v}{a_{0N}}). \qquad (1.3.2)$$

We can write:

$$C_D(v/a_{0N}) = C_D(M),$$

where $C_D = C_D(M)$ are the tabulated values of the drag coefficient as a function of Mach number.

Thus, the G-function (1.2.9) is transformed into a function of the projectile speed v expressed by the relation (1.3.2).

For example, we can consider that the drag coefficient $C_D(v/a)$ is obtained experimentally in a wind tunnel at the sea level, where the speed of sound is $a_{0N} = 341.458 m/s$, the air density is $\rho_{0N} = 1.2034 kg/m^3$, the air pressure is $p_{0N} = 750 mmHg.$, and the virtual temperature of air is $\tau_{0N} = 290K$.

For such model, the drag coefficient of the standard projectile is $C_D = C_D(v/a_{0N})$ while the G-function (1.2.9) in the ASM atmosphere, determined by (1.2.14), is

$$G_D(v) = 4.726 \cdot 10^{-4} \cdot v^2 C_D(v/a_{0N}), \qquad (1.3.3)$$

where $a_{0N} = 341.458 m/s$.

Making use of (1.3.2), for the drag acceleration (1.2.8), we can write:

$$a_D = c \cdot h(y) \cdot (3.927 \times 10^{-4} \rho_{0N}) v^2 C_D(\frac{v}{a_{0N}}), \qquad (1.3.4)$$

where the density function is

$$h(y) = (\frac{\tau_{0N} - 0.006328y}{\tau_{0N}})^{4.4} . \qquad (1.3.5)$$

The acceleration of the projectile (1.3.4) can be written as:

$$a_D = c \cdot h(y) \cdot G_D(v). \qquad (1.3.6)$$

The vector differential equation of the projectile flight (1.2.12) is

$$\frac{d\vec{v}}{dt} = \vec{g} - c \cdot h(y) \cdot (3.927 \times 10^{-4} \rho_{0N}) v^2 C_D (v / a_{0N}) \frac{\vec{v}}{v} , \qquad (1.3.7)$$

where

$$h(y) = (\frac{\tau_{0N} - 0.006328y}{\tau_{0N}})^{4.4} , \qquad c = \frac{id^2}{m} 1000 . \qquad (1.3.8)$$

The drag acceleration (1.3.4) for a given atmosphere (not necessary ASM, ICAO or TSA) is

$$a_D = c \cdot h_0(y) \cdot (3.927 \times 10^{-4} \rho_0) v^2 C_D(\frac{v}{a_0}) \qquad (1.3.9)$$

where

$$h_0(y) = (\frac{\tau_0 - 0.006328y}{\tau_0})^{4.4} \qquad (1.3.10)$$

is the density function.

At the sea level the density, the virtual temperature, the pressure, and the speed of sound are respectively ρ_0, τ_0, p_0 and a_0.

1.4. G-Functions of Resistance in ASM Atmosphere

In the contemporary literature, the differential equations that describe the flight of a projectile contain the function of resistance

$G_D(v)$, or the function of the drag coefficient $C_D(v/a)$, that are usually given in tabular forms.

Nowadays, the modern PC programs use successfully the tabulated values of the drag coefficient $C_D(v/a)$ to solve numerically the differential equations of the projectile flight for any projectile.

Thus, Robert McCoy uses the drag coefficient tables of some different types of standard projectiles to calculate trajectories of the projectiles. [7]

But, the tabulated values of a drag function $C_D(v/a)$, or a tabulated function of resistance $G_D(v)$, do not give the possibility to have analytical solutions for the differential equations of the projectile flight.

Even the Siacci's solution of the differential equations of the projectile flight that makes use of the tabulated drag coefficients $C_D(v/a)$ is a set of four large tabulated primary Siacci's functions that takes a lot of pages of any "Exterior Ballistics" book. [8]

In the "*Exterior Ballistics with Applications*" and in the "*Exterior Ballistics of Small Arms*," we use a simple and compact analytical Siacci function of resistance, [9],

$$K_D(v) = \begin{cases} 1.212 \cdot 10^{-4} v^2 & for \quad v \le 256 m/s \\ (v-240)/3 & for \quad v > 256 m/s \end{cases}, \qquad (1.4.1)$$

which enables us to obtain the three Siacci's primary analytical functions and the fourth one as an integral that can be easily calculated using a graphing calculator, or other PC techniques.

---

7. McCoy, Robert, "Modern Exterior Ballistics", page 112, Schiffer Publishing, 1999.
8. See for example the Siacci's functions of McCoy, R. "Modern Exterior Ballistics", chapter 6, Schiffer publishing, 1999..
9. Klimi, G, "Exterior Ballistics with Applications", chapter 5, Xlibris, 2009.

As we show hereafter, using the definition (1.3.3) of the G-function for the ASM standard atmosphere ($\rho_{0N} = 1.2034 kg / m^3$), i.e., using

$$G_D(v) = 4.726 \cdot 10^{-4} \cdot v^2 C_D(v / a_{0N}), \qquad (1.4.2)$$

for $a_{0N} = 341.458 m / s$ and the tabulated values of the drag coefficient $C_D = C_D(v / a_{0N})$, it is possible to obtain simple analytical G-functions of resistances.

Considering the relationship (1.4.2) between $G_D(v)$ and $C_D(v / a_{0N})$, as well as the idea that emerges from the analytical formula (1.4.1), using the tables of the standard drag coefficients stored in McCoy's PC program Mctraj41.Bas, or the tables of H.A.N.S. Cronander presented in *"Standard Drag Coefficients"*[10], we are able to obtain analytical formulas for the function of resistance $G_D(v)$ related with the corresponding drag coefficient function, $C_D(v / a_{0N})$, that is used by the modern exterior ballistics.

We consider that

$$C_D(v / a_{0N}) = C_D(M), \qquad (1.4.3)$$

where $C_D = C_D(M)$ are the tabulated values of the drag coefficient as function of the Mach number.

Standard Drag Coefficient Functions, $C_D(v / a_{0N})$

We denote:

$$C_{D1} = C_{D1}(v / a_{0N}), \; C_{D2} = C_{D2}(v / a_{0N}),$$
$$C_{D5} = C_{D5}(v / a_{0N}), \; C_{D6} = C_{D6}(v / a_{0N}),$$
$$C_{D7} = C_{D7}(v / a_{0N}), \; C_{D8} = C_{D8}(v / a_{0N}),$$
$$C_{DI} = C_{DI}(v / a_{0N}),$$

---

10. Hampus Anders Njord Sindre Cronander,"G-Dragfunctions.xls; G-Dragmodels.xls, http://www.cronander.net/, November 10th, 2005.

The Equations of Projectile Flight

respectively the drag coefficients of type 1, type 2, type 5, type 6, type 7, type 8, and Ingalls' projectiles.

Standard Function of Resistance, $G_D(v)$

We denote:

$$G_1 = G_1(v), G_2 = G_2(v),\ G_5 = G_5(v),\ G_6 = G_6(v),$$
$$G_7 = G_7(v),\ G_8 = G_8(v),$$
$$\text{and the Ingalls } G_I = G_I(v),$$

the G-function of resistance that correspond respectively to the drag coefficient function $C_D = C_D(v / a_{0N})$, based on the relationship (1.4.2).

For demonstration of the method we use to obtain a (two) piecewise G-function, in table 1 it is shown the drag coefficient of the standard projectile type 1, $C_{D1}(v / a_{0N})$, as a function of Mach number.

Table 1: Standard Drag coefficient $C_M(v / a)$ for the G_1-Type standard projectile

$M = v / a$	0.05	0.10	0.20	0.30	0.40	0.50	0.60	0.70
$C_M(v/a)$	0.2558	0.2487	0.2344	0.2213	0.2105	0.2035	0.2038	0.2168
$M = v / a$	0.75	0.80	0.85	0.90	0.95	1.00	1.05	1.10
$C_M(v/a)$	0.2308	0.2551	0.2913	0.3414	0.4072	0.4832	0.5429	0.5876
$M = v / a$	1.15	1.20	1.25	1.30	1.40	1.50	1.60	1.70
$C_M(v/a)$	0.6194	0.6406	0.6524	0.6587	0.6623	0.6571	0.6472	0.6339
$M = v / a$	1.80	1.90	2.00	2.10	2.20	2.30	2.40	2.50
$C_M(v/a)$	0.6202	0.6068	0.5937	0.5811	0.5692	0.5582	0.5481	0.5393
$M = v / a$	2.60	2.80	3.00	3.50	4.00	4.50	5.00	
$C_M(v/a)$	0.5318	0.5212	0.5134	0.5041	0.5005	0.4994	0.4988	

Using the relation (1.4.2) and the data given in the table 1, we obtain the corresponding table 2 in which are displayed the values of the functions $G_1 = G_1(v)$ defined by (1.4.2), as a function of projectile speed $v = a_{0N} M$ (where $a_{0N} = 341.458 m / s$).

Table 2: $G_1 = G_1(v)$ Function for the G1-Type standard projectile

	$G_1(v) = 4.726 \times 10^{-4} \cdot v^2 C_{D1}(v/a_{0N})$ $a_{0N} = 341.458 m/s$, $\rho_{0N} = 1.2034 kg/m^3$							
v	17.0729	34.1458	68.2916	102.4374	136.5832	170.7290	204.8748	239.0206
$G_1 = G_1(v)$	0.04	0.14	0.5166	1.0975	1.8558	2.8033	4.0427	5.8536
v	256.09	273.17	290.239	307.3122	324.3851	341.458	358.531	375.6038
$G_1 = G_1(v)$	7.15	9.00	11.597	15.2376	20.2499	26.6253	32.9812	39.1774
v	392.68	409.75	426.823	443.8954	478.0412	512.187	546.333	580.4786
$G_1 = G_1(v)$	45.14	50.83	56.1697	61.3398	71.5285	81.4671	91.295	100.9455
v	614.62	648.77	682.916	717.0618	751.2076	785.353	819.499	853.645
$G_1 = G_1(v)$	110.72	120.70	130.857	141.2075	151.8024	162.71	173.961	185.7286
v	887.79	956.08	1024.37	1195.103	1365.832	1536.56	1707.29	
$G_1 = G_1(v)$	198.09	225.16	254.605	340.268	441.2578	557.24	687.123	

Employing the data presented in table 2, and the techniques of regression analysis, we have obtained the following analytical relation for the function of resistance $G_1 = G_1(v)$:

$$G_1(v) = \begin{cases} 1.7152 \times 10^{-4} \cdot v^2 & for \quad v \le 256 m/s \\ 0.31123v - 77.8162 & for \quad 256 < v \le 1000 \end{cases} \qquad (1.4.4)$$

Note that the $G_1 = G_1(v)$ function obtained above is an approximate analytical function and reflects the accuracy of the data of the drag coefficient $C_M(v/a)$ displayed in table 1.

G-Functions of Resistance, ASM Atmosphere

The standard functions of resistance $G_D(v)$ that are shown below are obtained using McCoy's data on standard drag coefficients stored in his PC program Mctraj41.Bas. The following $G_D(v)$ functions of resistance are valid for the ASM atmosphere.

Note that for the same standard projectile type, hereafter we have provided at least one G-functions. The G-function that represents the better approximation is the one that belongs to the smallest interval of the projectile speed.

Anyway, for the same type of the projectile, we can use each one of G-functions and have practically the same results as far as the form coefficient (or the ballistics coefficient) is obtained using the respective G-function.

In the ASM standard atmosphere (the air density at the sea level is $\rho_{0N} = 1.2034 kg / m^3$), the G-function is

$$G_D(v) = (4.726 \times 10^{-4})v^2 C_D(\frac{v}{a_{0N}}),$$

where $a_{0N} = 341.458 m / s$ is the speed of sound at the sea level.

Using the method we have shown for the G_1-function, we have obtained the following G-functions for the ASM standard atmosphere:

Standard Function of Resistance, $G_1 = G_1(v)$, Type 1 Projectiles

$$G_1(v) = \begin{cases} 1.00347 \times 10^{-4} \cdot v^2 & for \quad v \leq 256 m/s \\ 0.312914v - 79.3976 & for \quad 256 < v \leq 1000 \end{cases}, \quad (1.4.5)$$

or

$$G_1(v) = \begin{cases} 1.00347 \times 10^{-4} \cdot v^2 & for \quad v \leq 256 m/s \\ 0.332044v - 87.6845 & for \quad 256 < v \leq 1200 \end{cases}. \quad (1.4.6)$$

Standard Function of Resistance, $G_2 = G_2(v)$, Type 2 Projectiles

$$G_2(v) = \begin{cases} 9.116233 \times 10^{-5} \cdot v^2 & for \quad v \leq 256 m/s \\ 0.141300 \cdot v - 29.9097 & for \quad 256 < v \leq 1300 m/s \end{cases}, \quad (1.4.7)$$

or

$$G_2(v) = \begin{cases} 9.116233 \times 10^{-5} \cdot v^2 & for \quad v \leq 256 m/s \\ 0.146467 \cdot v - 32.7991 & for \quad 256 < v \leq 1700 m/s \end{cases}. \quad (1.4.8)$$

Standard Function of Resistance, $G_5 = G_5(v)$, Type 5 Projectiles

$$G_5(v) = \begin{cases} 7.43571 \times 10^{-5} \cdot v^2 & for \quad v \leq 256 m/s \\ 0.19734 \cdot v - 49.0806 & for \quad 256 < v \leq 1500 m/s \end{cases}, \quad (1.4.9)$$

$$G_5(v) = \begin{cases} 7.43571 \times 10^{-5} \cdot v^2 & for & v \leq 256 m/s \\ 0.203422 \cdot v - 52.6662 & for & 256 < v \leq 1700 m/s \end{cases}. \quad (1.4.10)$$

Standard Function of Resistance, $G_6 = G_6(v)$, Type 6 Projectiles

$$G_6(v) = \begin{cases} 1.0334 \times 10^{-4} \cdot v^2 & for & v \leq 256 m/s \\ 0.140533 \cdot v - 23.6633 & for & 256 < v \leq 1700 m/s \end{cases}. \quad (1.4.11)$$

Standard Function of Resistance, $G_7 = G_7(v)$, Type 7 Projectiles

$$G_7(v) = \begin{cases} 5.66480 \times 10^{-5} \cdot v^2 & for & v \leq 256 m/s \\ 0.150355 \cdot v - 34.7319 & for & 256 m/s < v \leq 1700 \end{cases}. \quad (1.4.12)$$

Standard Function of Resistance, $G_8 = G_8(v)$, Type 8 Projectiles

$$G_8(v) = \begin{cases} 9.9366 \times 10^{-5} \cdot v^2 & for & v \leq 256 m/s \\ 0.14733 \cdot v - 29.0850 & for & 256 < v \leq 1500 m/s \end{cases}, \quad (1.4.13)$$

or

$$G_8(v) = \begin{cases} 9.9366 \times 10^{-5} \cdot v^2 & for & v \leq 256 m/s \\ 0.150101 \cdot v - 30.6672 & for & 256 < v \leq 1700 m/s \end{cases}. \quad (1.4.14)$$

The Ingall's G-Function of Resistance

$$G_I(v) = \begin{cases} 1.08774 \times 10^{-4} \cdot v^2 & for & v \leq 256 m/s \\ 0.311318v - 78.932 & for & 256 < v \leq 800 m/s \end{cases}, \quad (1.4.15)$$

$$G_I(v) = \begin{cases} 1.08774 \times 10^{-4} \cdot v^2 & for & v \leq 256 m/s \\ 0.32072 \cdot v - 82.6909 & for & 256 < v \leq 1000 m/s \end{cases}, \quad (1.4.16)$$

or

$$G_I(v) = \begin{cases} 1.08774 \times 10^{-4} \cdot v^2 & for & v \leq 256 m/s \\ 0.3321 \cdot v - 86.0837 & for & 256 < v \leq 1200 m/s \end{cases}. \quad (1.4.17)$$

The Spherical Projectile G-Function of Resistance, $v \leq 1400 m/s$.

$$G_S(v) = 2.60444956 \times 10^{-4} v^2. \quad (1.4.18)$$

The Cylindrical Projectile G-Function of Resistance,
$150 < v \le 1200 m/s$

$$G_C(v) = 2.5117 \times 10^{-4} v^2 . \qquad (1.4.19)$$

The function of air density for the ASM standard atmosphere is

$$h(y) = (\frac{289.6 - 0.006328y}{289.6})^{4.4} . \qquad (1.4.20)$$

Notes:

- The form coefficient i of a given projectile and the corresponding ballistics coefficient,

$$c = \frac{id^2}{m} 1000 , \qquad (1.4.21)$$

should match the G-function of resistance that is used to determine them.

- If the form coefficient i is known, then to find the ballistics coefficient c, we simply substitute i in (1.4.21).
- If the ballistics coefficient is obtained using the formula:

$$C = \frac{m}{id^2} , \qquad (1.4.22)$$

and is expressed in English units (pound per square inch), then to find the ballistics coefficient that corresponds to formula (1.4.21), we can use the following relation:

$$C = \frac{1.4222}{c} . \qquad (1.4.23)$$

1.5. G-Functions of Resistance in ICAO Atmosphere

The G-functions related with the ICAO standard atmosphere are slightly different from the G-functions related with the ASM presented in section 1.4.

For the ICAO standard atmosphere, the G-function defined by the equation (1.2.17), is

$$G_{DA}(v) = 4.811 \times 10^{-4} \cdot v^2 C_D(v / a_{0N}),\qquad (1.5.1)$$

where the speed of sound and the air density at the sea level are respectively:

$$a_{0N} = 340.30m / s, \quad \rho_{0N} = 1.2251kg / m^3.\qquad (1.5.2)$$

The density function is

$$h(y) = (\frac{288.15 - 0.006328y}{288.15})^{4.4}.\qquad (1.5.3)$$

As it was demonstrated for the Mayevski's G-function (section 1.6), using (1.5.1) we have found the following G-functions for the ICAO atmosphere:

Standard Function of Resistance, $G_{1A} = G_{1A}(v)$, Type 1 Projectiles

$$G_{1A}(v) = \begin{cases} 1.0584 \times 10^{-4} \cdot v^2 & for \quad v \le 256m / s \\ 0.315754 \cdot v - 78.6769 & for \quad 256 < v \le 1000 \end{cases},\qquad (1.5.4)$$

or

$$G_{1A}(v) = \begin{cases} 1.0584 \times 10^{-4} \cdot v^2 & for \quad v \le 256m / s \\ 0.331547v - 86.0227 & for \quad 256 < v \le 1200 \end{cases}.\qquad (1.5.5)$$

Standard Function of Resistance, $G_{2A} = G_{2A}(v)$, Type 2 Projectiles

$$G_{2A}(v) = \begin{cases} 9.2868 \times 10^{-5} \cdot v^2 & for \quad v \le 256m / s \\ 0.143353 \cdot v - 30.2415 & for \quad 256 < v \le 1300m / s \end{cases},\qquad (1.5.6)$$

or

$$G_{2A}(v) = \begin{cases} 9.2868 \times 10^{-5} \cdot v^2 & for \quad v \le 256m / s \\ 0.148718 \cdot v - 33.2882 & for \quad 256 < v \le 1700m / s \end{cases}.\qquad (1.5.7)$$

Standard Function of Resistance, $G_{5A} = G_{5A}(v)$, Type 5 Projectiles

$$G_{5A}(v) = \begin{cases} 7.5244 \times 10^{-5} \cdot v^2 & for & v \leq 256 m/s \\ 0.200207 \cdot v - 49.625 & for & 256 < v \leq 1500 m/s \end{cases}, \qquad (1.5.8)$$

or

$$G_{5A}(v) = \begin{cases} 7.5244 \times 10^{-5} \cdot v^2 & for & v \leq 256 m/s \\ 0.206378 \cdot v - 53.2504 & for & 256 < v \leq 1700 m/s \end{cases}. \qquad (1.5.9)$$

Standard Function of Resistance, $G_{6A} = G_{6A}(v)$, Type 6 Projectiles

$$G_{6A}(v) = \begin{cases} 1.05244 \times 10^{-4} \cdot v^2 & for & v \leq 256 m/s \\ 0.142352v - 23.6937 & for & 256 < v \leq 1700 m/s \end{cases}. \qquad (1.5.10)$$

Standard Function of Resistance, $G_{7A} = G_{7A}(v)$, Type 7 Projectiles

$$G_{7A}(v) = \begin{cases} 5.7679 \times 10^{-5} \cdot v^2 & for & v \leq 256 m/s \\ 0.152593 \cdot v - 35.1717 & for & 256 < v \leq 1700 m/s \end{cases}. \qquad (1.5.11)$$

Standard Function of Resistance, $G_{8A} = G_{8A}(v)$, Type 8 Projectiles

$$G_{8A}(v) = \begin{cases} 1.01154 \times 10^{-4} \cdot v^2 & for & v \leq 256 m/s \\ 0.149441 \cdot v - 29.3790 & for & 256 < v \leq 1500 m/s \end{cases}, \qquad (1.5.12)$$

or

$$G_{8A}(v) = \begin{cases} 1.01154 \times 10^{-4} \cdot v^2 & for & v \leq 256 m/s \\ 0.152307 \cdot v - 31.0336 & for & 256 < v \leq 1700 m/s \end{cases}. \qquad (1.5.16)$$

The Ingall's G-Function of Resistance, $G_{IA} = G_{IA}(v)$

$$G_{IA}(v) = \begin{cases} 1.0724 \times 10^{-4} \cdot v^2 & for & v \leq 256 m/s \\ 0.315181 \cdot v - 79.1941 & for & 256 < v \leq 800 m/s \end{cases}, \qquad (1.5.17)$$

$$G_{IA}(v) = \begin{cases} 1.0724 \times 10^{-4} \cdot v^2 & for & v \leq 256 m/s \\ 0.325383 \cdot v - 83.6082 & for & 256 < v \leq 1000 m/s \end{cases}, \qquad (1.5.18)$$

or

$$G_{IA}(v) = \begin{cases} 1.0724 \times 10^{-4} \cdot v^2 & for & v \le 256m/s \\ 0.336927v - 89.0608 & for & 256 < v \le 1200m/s \end{cases}. \quad (1.5.19)$$

The Spherical Projectile G-Function of Resistance, $G_{SA} = G_{SA}(v)$

$$G_{SA}(v) = 2.7189 \times 10^{-4} v^2, \text{ for } v \le 1400m/s. \quad (1.5.20)$$

The Cylindrical Projectile G-Function of Resistance, $G_{CA} = G_{CA}(v)$

$$G_{CA}(u) = 2.55078 \times 10^{-4} v^2, \text{ for } 150 < v \le 1200m/s. \quad (1.5.21)$$

1.6. G-Functions of Resistance in TSA Atmosphere

Mayevski's Approximate G-Function of Resistance

Mayevski's analytical G-function of resistance is derived employing the data obtained in Krupp's experiments for a given standard projectile.

The calculations of most of the trajectories of different projectiles and the compilation of respective ballistics tables for at least 80 years, since around 1883, are based on the Mayevski's function of resistance.

Since the Mayevski's function of resistance is still with interest in exterior ballistics, not to mention the historical importance for some generation of ballisticians and its use in the ballistics literature, we have approximated the Mayevski's function of resistance with a simple (two) piece wise analytical function of resistance, (formula (1.6.5)).

The Mayevski-Zabudski's drag coefficient $C_M(v/a)$ for Mayevski's standard projectile is given in the following table 1.[11]

---

11. Edoardo Mori, "Balistica teorica e pratica"; "http://www.earmi.it/balistica/formi.htm"

The Equations of Projectile Flight

Table 1:—Drag coefficient $C_M(v/a)$ for the Mayevski's standard projectile

$M = v/a$	0.1	0.2	0.3	0.4	0.5	0.6	0.7	0.8
$C_M(v/a)$	0.228	0.228	0.228	0.228	0.228	0.228	0.228	0.259
$M = v/a$	0.9	1.0	1.1	1.2	1.3	1.4	1.5	1.6
$C_M(v/a)$	0.313	0.430	0.573	0.626	0.643	0.643	0.643	0.643
$M = v/a$	1.7	1.8	1.9	2	2.1	2.2	2.3	2.4
$C_M(v/a)$	0.633	0.623	0.613	0.603	0.594	0.586	0.578	0.571
$M = v/a$	2.5	2.6	2.7	2.8	2.9			
$C_M(v/a)$	0.559	0.549	0.540	0.531	0.521			

Making use of the Mayevski's G-function of resistance (for $a_{0N} = 340.83 m/s$),

$$G_M(v) = 4.732 \cdot 10^{-4} \cdot v^2 C_M(v/a_{0N}), \tag{1.6.1}$$

and the drag coefficient $C_M = C_M(v/a_{0N})$ presented in table 1, we have obtained the following expression for the Mayevski's G-function of resistance:

$$G_D(v) = 0.320243 \cdot v - 81.3721, \tag{1.6.2}$$

for projectile speed $v > 256 m/s$ till around $v = 900 m/s$.

For the speed of projectile $v \le 256 m/s$, we have considered the original Mayevski's G-function (ref. Table 1, page 83, "*Exterior Ballistics with Applications*"),

$$G_D(v) = 1.0807 \times 10^{-4} \cdot v^2. \tag{1.6.3}$$

Therefore, the approximate Mayevski's G-function is

$$G_M(v) = \begin{cases} 1.0807 \times 10^{-4} \cdot v^2 & \text{for } v \le 256 m/s \\ 0.320243 \cdot v - 81.3721 & \text{for } v > 256 m/s \end{cases} \tag{1.6.4}$$

The Russian G-function of Year 1943

Another G-function in use by the Russian army is the so called law of resistance of 1943. It is given in a tabular form.

In analytical form, the Russian law of year 1943 can be presented approximately using the following two-piecewise function:

$$G_{43}(v) = \begin{cases} 7.454 \times 10^{-5} \cdot v^2 & for \quad v \leq 256m/s \\ 0.157713 \cdot v - 36.39542 & for \quad v > 256m/s \end{cases} \qquad (1.6.5)$$

For speeds less than 256m/s, the coefficient in front of v^2 is not constant, but it is considered approximately equal to the value 7.454×10^{-5} of the G_{43}-function that corresponds to the projectile speed 100m/s.

Approximate Siacci's G-function

In this group of G-functions, we will include also the Siacci G-function [12]:

$$K_D(v) = \begin{cases} 1.212 \cdot 10^{-4} v^2 & for \quad v \leq 256m/s \\ (v - 240)/3 & for \quad v > 256m/s \end{cases}. \qquad (1.6.6)$$

We consider that the Mayevski's G-function, the G_{43}-function and the Siacci G-functions are related with the Traditional Standard Atmosphere (TSA) in which the speed of sound at the firing point (located at the sea level) is $a_{0N} = 340.83m/s$, the relative humidity of air and the virtual temperature are respectively 50% and $\tau_{0N} = 289.08$ Kelvin, while the density of air is $\rho_{0N} = 1.205kg/m^3$ (ref. "Exterior Ballistics with Applications, p.121).

The density function is

$$h(y) = (\frac{289.08 - 0.006328y}{289.08})^{4.4}. \qquad (1.6.7)$$

---

12. Klimi, G., "Exterior Ballistics with Applications", p. 71, Xlibris, 2008

1.7. General Form of the G-function

Any G-function of resistance (except the spherical and cylindrical G-function) in ASM or ICAO atmosphere has the general form:

$$G_D(v) = \begin{cases} A \cdot v^2 & for \quad v \le 256 m/s \\ E \cdot v - F & for \quad v > 256 m/s \end{cases}, \qquad (1.7.1)$$

where the values of the parameters A, E, and F depend on the G-function in use, as are shown in sections 1.4, 1.5 and 1.6.

For example, the values of A, E, and F for the G_1-function in ASM atmosphere are respectively (see formula 1.4.5):

$$A = 1.00347 \times 10^{-4}, \qquad E = 0.312914, \qquad F = 79.3976.$$

We can use the general form (1.7.1) of the G-function to integrate analytically the differential equation of the projectile flight, and then using the respective values of the parameters A, E, and F, we find the trajectory of the projectile flight.

The vector differential equations of the projectile flight (1.3.7) can be written:

$$\frac{d\vec{v}}{dt} = \vec{g} - c \cdot h(y) \cdot G_D(v) \frac{\vec{v}}{v}, \qquad (1.7.2)$$

where

$$h(y) = (\frac{\tau_{0N} - 0.006328y}{\tau_{0N}})^{4.4}, \qquad (1.7.3)$$

while $G_D(v)$ is the general form of G-function (1.7.1) of a given standard projectile.

2

PROJECTILE TRAJECTORY

To describe the ballistics trajectory of a projectile, we use the system of differential equations obtained from the vector differential equation (1.7.2).

We show four universal PC programs that enable us to solve the differential equations of projectile flight and obtain answers for the main problems of exterior ballistics.

In addition to the solution of exterior ballistics problems using the PC programs, we introduce as well *the projectile trajectory-streamline and Snell's law model*, demonstrating at the same time the truthfulness and accurateness of the original model in the solution of exterior ballistics problems.

The assumption of "wave behavior" of the projectiles and Newton's interpretation of Snell's law applied to the motion of projectile give remarkably correct results for the solution of problems of exterior ballistics.

The solution of differential equations that describe the flight of a point-mass projectile is related with the ballistics coefficient (BC) that is an important ballistics characteristics of a given projectile and firearm related with a given G-drag function.

Though BC is defined as a constant,

$$c = \frac{id^2}{m} 1000 \,,$$

actually it is a function of the projectile speed.

In practice, the ballistics coefficient is obtained performing firing tests for a set of ranges. The BC obtained as result of experiments, corresponds to a given G-function of resistance, and is a function of the departure angle of the projectile, $c = f(\alpha_0)$, presented in tabular form.

To determine experimentally the BC, and construct the table of ballistics coefficients, it is recommended that:

- for artillery firearms, the firing tests to determine the BC function should be performed for the set of departure angles: 5, 15, 25, 35, and 45 degrees, or for the set of angles: 5, 10, 20, 30, 40, 60 degrees (the largest angle depends on the maximum departure angle);
- for the artillery firearms of a caliber less than 75mm, the BC function can be determined by shooting test for four departure angles;
- for small arms, the ballistics coefficients should be measured for horizontal ranges: 500m, 800m, 1,500m, 2,000m, 3,000m and 4,000m.

For intermediate angles, the ballistics coefficient is determined by interpolation.

2.1. Differential Equations of the Projectile Trajectory

The motion of a point-mass projectile (with respect to a rectangular Cartesian system of coordinates) in presence of gravity and the drag force is described by the vector differential equation of the projectile flight (1.7.2), i.e., by the equation:

$$\frac{d\vec{v}}{dt} = \vec{g} - c \cdot h(y) \cdot G_D(v)\frac{\vec{v}}{v}, \qquad (2.1.1)$$

where

$$G_D(v) = \begin{cases} A \cdot v^2 & for \quad v \le 256m/s \\ E \cdot v - F & for \quad v > 256m/s \end{cases}, \qquad (2.1.2)$$

$$h(y) = (\frac{\tau_{0N} - 0.006328y}{\tau_{0N}})^{4.4} , \qquad c = \frac{id^2}{m} 1000 .$$

For a particular type of G-function, G_1, G_2, etc., the values of the parameters A, E, and F in G-function (2.1.2) are given in section 1.4 and section 1.5.

The vector differential equation (2.1.1) can be used to determine the projectile trajectory and to solve the problems of exterior ballistics in ASM, ICAO, or TSA atmosphere.

From the vector differential equation (2.1.1), we obtain the following systems of differential equations that describe the projectile trajectory in standard atmosphere (the ballistics characteristics of the projectile and firearm are supposed to be standard):
Variable t (time)

$$\begin{cases} \dfrac{dv_x}{dt} = -c \cdot h(y) \cdot G_D(v) \cdot \dfrac{v_x}{v} \\[2mm] \dfrac{dp}{dt} = -\dfrac{g}{v_x} \\[2mm] \dfrac{dx}{dt} = v_x \\[2mm] \dfrac{dy}{dt} = v_y \end{cases} ; \qquad (2.1.3)$$

Variable x (abscissa)

$$\begin{cases} \dfrac{dv_x}{dx} = -c \cdot h(y) \cdot \dfrac{G_D(v)}{v} \\[2mm] \dfrac{dp}{dx} = -\dfrac{g}{v_x^2} \\[2mm] \dfrac{dt}{dx} = \dfrac{1}{v_x} \\[2mm] \dfrac{dy}{dx} = p \end{cases} , \qquad (p = \tan\alpha) \quad (2.1.4)$$

Variable $p = \tan \alpha$, (α -angle of flight)

$$\begin{cases} \dfrac{dv_x}{dp} = \dfrac{c \cdot h(y)}{g} G_D(v) \dfrac{v_x}{(1+p^2)^{1/2}} \\[2mm] \dfrac{dt}{dp} = -v_x / g \\[2mm] \dfrac{dx}{dp} = -v_x^2 / g \\[2mm] \dfrac{dy}{dp} = -pv_x^2 / g \end{cases} \qquad , \qquad (2.1.5)$$

where

$$h(y) = (\dfrac{\tau_{0N} - 0.006328y}{\tau_{0N}})^{4.4}, \qquad (2.1.6)$$

and

$$c = \dfrac{id^2}{m} 1000. \qquad (2.1.7)$$

For the ASM atmosphere, the standard virtual temperature at the sea level is $\tau_{0N} = 290K$. For the ICAO atmosphere, the virtual temperature at the sea level is $\tau_{0N} = 288.15K$. For the TSA atmosphere, $\tau_{0N} = 289.08K$.

The above systems can be integrated numerically to obtain the elements of the projectile trajectory at any point over or down the horizontal level (x-axis).

In practice, in case the ballistics density ρ_B is known (for example, measured by a balloon-borne radiosonde), we substitute in the system of equations (2.1.4), (2.1.5), and (2.1.6) the density function $h(y)$ with ρ_B / ρ_{0N}.

Ballistics density is a hypothetical value ρ_B of the density of air whose effect on the projectile trajectory is the same as the effect of the density of air the projectile encounters during the flight.

Example 1.1

Write the system of differential equation (2.1.4) for a projectile fired in the ICAO atmosphere with a speed of 850m/s at an angle of 5 degrees if the ballistics coefficient that corresponds to the G_5 function is 4.799m²/kg. The firing site is at the sea level.

Assume that the speed of the projectile during the flight is greater than 256m/s.

Solution:

The G_5 function that corresponds to the ICAO atmosphere, given by (1.5.8), is

$$G_{5A}(v) = \begin{cases} 7.5244 \times 10^{-5} \cdot v^2 & for & v \leq 256m/s \\ 0.200207 \cdot v - 49.625 & for & 256 < v \leq 1500m/s \end{cases}.$$

Substituting in (2.1.4) the second part of the G_5-function, we obtain the system of differential equations

$$\begin{cases} \dfrac{dv_x}{dx} = -4.799 \cdot (\dfrac{288.15 - 0.006328y}{288.15})^{4.4} \dfrac{0.200207v_x(1+p^2)^{1/2} - 49.625}{v_x(1+p^2)^{1/2}} \\ \\ \dfrac{dp}{dx} = -\dfrac{g}{v_x^2} \\ \dfrac{dt}{dx} = \dfrac{1}{v_x} \\ \dfrac{dy}{dx} = p \end{cases},$$

that we can use to solve the ballistics problems for the given projectile flying in the ICAO atmosphere. The initial conditions (at the departure point) are:

$$x_0 = 0, \; y_0 = 0, \; v_0 = 850m/s, \; \alpha_0 = 5°.$$

The above system can be solved using methods of numerical integration.

2.2. Characteristics of Atmosphere

In general, the characteristics of the atmosphere at the firing site are different from the characteristics of the atmosphere at the sea level.

The characteristics of atmosphere are measured by the metro-stations that usually are not at the firing site. A metro-station bulletin for artillery fire gives as well the ballistics temperature, ballistics density, ballistics pressure, ballistics wind and its direction that are important data for the artillery shooting and fire corrections.

In absence of the data from metro-stations, we can measure directly the atmospheric pressure, the temperature of dry air, and the relative humidity of atmospheric air.

In both situations, we have to estimate the necessary characteristics of atmosphere at firing site, based on the measurements.

The following ways to estimate the characteristics of atmosphere are based on the formulas presented in chapter eight of *"Exterior Ballistics with Applications"*.

Virtual Temperature, Density, and the Speed of Sound

The projectile flight is influenced by the characteristics of the atmosphere (pressure p_0, density ρ_0, virtual temperature τ_0, speed of sound a_0) at the firing site and by the changes of those characteristics with the altitude of the projectile, represented by the density function

$$h_0(y) = (\frac{\tau_0 - 0.006328y}{\tau_0})^{4.4} .$$

We assume that the virtual temperature changes with altitude over the firing site according to the law

$$\tau = \tau_0 - 0.006328y .$$

Since the virtual temperature depends as well on the air humidity, we have to correct the temperature of air, considering the humidity at the firing site. The increase of humidity decreases the density of the atmospheric air, and as result, the range of the projectile increases.

Using the measured pressure, the temperature of dry air, and the humidity of atmospheric air, we are able to find the virtual temperature and all other characteristics of the atmosphere at the measuring location (at the metro-station located at the sea level, or at the shooting site).

For a given relative humidity, the virtual temperature at the firing site is

$$\tau_0 = \frac{T_0}{(1 - 0.3785 \cdot e_0 / p_0)}, \qquad (2.2.1)$$

where T_0 is the temperature of dry air, p_0 is the atmospheric pressure, and e_0 is the pressure of water vapor that corresponds to the given relative humidity.

In (2.2.1), the temperature is in Kelvin, while pressure is in mm Hg.

To estimate the pressure of water vapor e, we can use table 1, page 91 of "*Exterior Ballistics of Small Arms*", (Klimi, G, Xlibris, 2009), i.e.:

Table 1:—Saturated Pressure of Water Vapor (Humidity 100%)

Temperature	-40	-18	-10	0.0	5.0	10	15
Pressure e	0.15	1.14	1.95	4.58	6.54	9.20	12.7
Temperature	20	25	30	38	40	54	-
Pressure e	17.54	23.76	31.7	49.2	55.1	115.1	-

The following examples illustrate the methods we use to estimate the virtual temperature and the pressure using the data presented in table 1.

Once the virtual temperature is calculated, we can find the air density and the corresponding speed of sound respectively using the following formulas:

$$\rho_0 = \frac{\mu_a \, p_0}{R \, \tau_0}, \qquad (2.2.2)$$

and

$$a_0 = 20.0469 \sqrt{\tau_0}, \qquad (2.2.3)$$

where

$\mu_a = 28.9644 kg/mol^{-1}$ is the molar mass of air, and
$R = 8.31441 Jmol^{-1}K^{-1}$ is the universal constant of gases.
    Substituting the above values in (2.2.2), the density can be written:

$$\rho_0 = 3.4836 \times 10^{-3} \frac{p_0}{\tau_0}. \tag{2.2.4}$$

Calculation of the Characteristics of Atmosphere at the Shooting Site

Using the data measured by the metro-station [13], we can use the following approach to find the characteristics of the atmosphere at the shooting site.

I. First Approach

The temperature of dry air, the density, and the pressure change with altitude over the sea level according to the already known laws:

$$T = T_0 - 0.006328y, \tag{2.2.5}$$

$$\rho = \rho_0 h_0(y) = \rho_0 \left(\frac{T_0 - 0.006328y}{T_0}\right)^{4.4}, \tag{2.2.6}$$

and

$$p = p_0 h_0(y) = p_0 \left(\frac{T_0 - 0.006328y}{T_0}\right)^{5.4}. \tag{2.2.7}$$

Using the data from the metro-station and the above formulas, first we estimate the corresponding data for the sea level, and then we find the respective data at the shooting site. We illustrate this approach in example 3.3.

II. Practical Rule to Estimate the Atmospheric Pressure

In practice, in absence of barometers, to estimate the atmospheric pressure at a given altitude, when it is known the atmospheric

---

13.   Refer to sections 8.2-8.3 of "Exterior Ballistics with Applications"

pressure at the sea level or at the altitude of meteorological station is known, we can use the following practical rule:

The atmospheric pressure decreases 8-9 mm Hg for any 100 meters increase in altitude (the smaller increase corresponds to very high altitudes; in the following formula, we use the average 8.5).

The pressure at altitude y can be estimated approximately by the formula:

$$p = p_0 - 8.5 \cdot (y - y_0) / 100, \qquad\qquad (2.2.8)$$

where p_0 is the pressure at the altitude y_0.

Thus, if the pressure p_0 at the sea level is 760mmHg, then at the altitude 100 meters over the sea level, it is 751mmHg, while at the altitude 1,000 meters, the pressure is approximately

$$p = p_0 - 8.5 \cdot (y - y_0) / 100 = 750 - 8.5 \cdot (1000 - 0) / 100 = 665mmHg.$$

III. Second Approach

The second approach is similar to the first one, but is based on the following formulas:

$$\tau = \tau_0 - 0.006328y, \qquad\qquad (2.2.9)$$

$$\rho = \rho_0 \left(\frac{\tau_0 - 0.006328y}{\tau_0} \right)^{4.4}, \qquad\qquad (2.2.10)$$

and

$$p = p_0 \left(\frac{\tau_0 - 0.006328y}{\tau_0} \right)^{5.4}. \qquad\qquad (2.2.11)$$

We illustrate the second approach in example 3.3.

For more information on the characteristics of the atmosphere, the reader should refer to the sections 8.1-8.3 of the manuscript book "*Exterior Ballistics with Applications*".

Example 2.1

The relative humidity of air in ASM atmosphere is 78%. Find the virtual temperature that corresponds to the standard temperature of 15 degrees. Atmospheric pressure is 750mmHg.

Solution:

The pressure of saturated water vapor (table 1) at 15 degrees Celsius is $e_{100\%} = 12.7mmHg$. The pressure of water vapor for the relative humidity 100% is

$$e_{15°;78\%} = 0.78 \cdot (e_{100\%}) = 0.78 \cdot (12.7) = 9.906mmHg.$$

The virtual temperature is

$$\tau = \frac{T}{1 - 0.3785(e/p)} = \frac{273.15 + 15}{1 - 0.3785(9.906/750)} = 289.6 \,°Kelvin.$$

Example 2.2

Find the virtual temperature at the firing site if:

(a) the pressure, the temperature of dry air, and the relative humidity are respectively 760mm, 25 degrees Celsius, and 65%;
(b) the temperature of dry air is 25 degrees Celsius, the pressure 750mm, and the humidity 65%;
(c) the temperature of dry air is 23 degree Celsius, the pressure 760mm and the humidity 65%.

Solution:

(a) The pressure of water vapors (table 1) is

$$e_{25°;65\%} = 0.65 \cdot (e_{100\%}) = 0.65 \cdot (23.76) = 15.444mmHg.$$

The virtual temperature is

$$\tau = \frac{T}{1 - 0.3785(e/p)} = \frac{298.15}{1 - 0.3785(15.444/760)} = 300.46 \,°Kelvin.$$

(b) The pressure of water vapor (table 1) is
   $e_{25°;65\%} = 15.444mmHg.$

The virtual temperature is

$$\tau = \frac{T}{1 - 0.3785(e/p)} = \frac{298.15}{1 - 0.3785(15.444/750)} = 300.49 \ °Kelvin$$

(c) By interpolation (see table 1), we find that at 23 degrees Celsius, the saturated vapor pressure is $e_{23°;100\%} = 21.272mmHg.$

The relative vapor pressure is

$$e_{23°;65\%} = 0.65 \cdot (e_{100\%}) = 0.65 \cdot (21.272) = 13.83mmHg.$$

The virtual temperature is

$$\tau = \frac{T}{1 - 0.3785(e/p)} = \frac{296.15}{1 - 0.3785(13.83/760)} = 298.20 \ °Kelvin.$$

Example 2.3

The temperature of dry air, the pressure, and the humidity at the sea level are respectively 15 degrees Celsius, 750mmHg, and 75%.

Find the virtual temperature at the altitude 1,000 meters over the sea level.

Solution:

Method 1

The temperature of dry air at the altitude 1,000 meters is

$$T = T_0 - 0.006328y = 288.15 - 0.006328 \cdot (1000) = 281.822 \ °K$$

Using (2.2.9), we find that the pressure is

$$p = p_0 \left(\frac{T_0 - 0.006328y}{T_0} \right)^{5.4} = 750\left(\frac{281.822}{288.15} \right)^{5.4} = 665.25mmHg$$

.

Note:

We find approximately the same value of air pressure using the approximate formula:

$$p = p_0 - 8.5(y - y_0)/100 = 750 - 8.5(1000 - 0)/100 = 665 mmHg.$$

Using table 1, we find that the pressure of water vapors is

$$e_{8.67°;75\%} = 6.37 mmHg.$$

The virtual temperature is

$$\tau = \frac{T}{1 - 0.3785(e/p)} = \frac{281.822}{1 - 0.3785(6.37/665.25)} = 282.85 \;°Kelvin.$$

Method 2

Using table 1, we find that the pressure of water vapor at the sea level is

$$e_{15°;75\%} = 9.525 mmHg.$$

The virtual temperature at the sea level is

$$\tau_0 = \frac{T_0}{1 - 0.3785(e_0/p_0)} = \frac{273.15 + 15}{1 - 0.3785(9.525/750)} = 289.54° K$$

The virtual temperature at the altitude 1,000 meters is

$$\tau = \tau_0 - 0.006328y = 289.54 - 0.006328 \cdot (1000) = 283.21 \;°K.$$

Note that both methods of calculations give approximately the same result.

Example 2.4

The temperature and the pressure of atmospheric air at altitude 1,500 meters are respectively 5.81 degrees Celsius, and 625.90 mmHg. The relative humidity is 50%.

Find the virtual temperature at the altitude 1,500 meters and at the sea level as well.

Solution:

Using the interpolation and the data of table 1, we find that the pressure of water vapors is $e_{5.81°;50\%} = 3.41mmHg..$

$$\tau = \frac{T}{1 - 0.3785(e/p)} = \frac{278.96}{1 - 0.3785(3.405/625.9)} = 279.53 \ °Kelvin$$

Employing

$$\tau = \tau_0 - 0.006328y,$$

for the virtual temperature at the sea level we find:

$$\tau_0 = \tau + 0.006328y = 279.53 + 0.006328(1500) = 289.022 \ °Kelvin.$$

2.3. Projectile Trajectory in Non-Standard Atmosphere

The differential equations, presented in section 2.1, describe the projectile trajectory in standard atmosphere. For a projectile that is launched in non-standard atmosphere, we can use any G-function of resistance determined for any standard atmosphere, for example the G-function determined for the ASM atmosphere, or the G-function determined for the ICAO atmosphere.

Indeed, let's consider an arbitrary chosen atmosphere (in general different from ASM, ICAO, or TSA) with the following characteristics at the sea level:

Density, ρ_0; pressure, p_0; virtual temperature, τ_0 that corresponds to a given relative humidity of air; speed of sound, a_0.

The temperature, the density, and the pressure change with altitude over the sea level according to the already known laws:

$$\tau = \tau_0 - 0.006328y,$$

$$\rho = \rho_0 h_0(y) = \rho_0 (\frac{\tau_0 - 0.006328y}{\tau_0})^{4.4},$$

$$p = p_0 h_0(y) = p_0 (\frac{\tau_0 - 0.006328y}{\tau_0})^{5.4}.$$

The drag acceleration (see formula 1.3.9) of the projectile flying in the given non standard atmosphere is

$$a_D = c \cdot h_0(y) \cdot (3.927 \times 10^{-4} \rho_0) v^2 C_D (\frac{v}{a_0}), \qquad (2.3.1)$$

where

$$h_0(y) = (\frac{\tau_0 - 0.006328y}{\tau_0})^{4.4}. \qquad (2.3.2)$$

is the density function.

In general, for a projectile, the drag coefficient $C_D(v/a_0)$ is unknown.

Let's assume that we know the G-function (1.4.2) defined for a given standard atmosphere, i.e., it is known:

$$G_D(v) = (3.927 \times 10^{-4} \rho_{0N}) v^2 C_D (\frac{v}{a_{0N}}). \qquad (2.3.3)$$

For example, for the ASM atmosphere: $a_{0N} = 341.458 m/s$, $\rho_{0N} = 1.2034 kg/m^3$.

The drag acceleration (2.3.1) of the given projectile can be written:

$$a_D = c \cdot \frac{\rho_0}{\rho_{0N}} \cdot h_0(y) \cdot (3.927 \times 10^{-4} \rho_{0N}) v^2 C_D (\frac{v}{a_0}), \qquad (2.3.4)$$

or,

$$a_D = c \cdot \frac{\rho_0}{\rho_{0N}} \cdot h_0(y) \cdot (3.927 \times 10^{-4} \rho_{0N}) v^2 C_D (\frac{v \cdot a_{0N}/a_0}{a_{0N}}). \qquad (2.3.5)$$

For the drag acceleration (2.3.5), we can write:

$$a_D = c \cdot \frac{p_0}{p_{0N}} \frac{a_0^2}{a_{0N}^2} \cdot h_0(y) \cdot (3.927 \times 10^{-4} \rho_{0N}) \cdot (v \cdot \frac{a_{0N}}{a_0})^2 C_D(\frac{v \cdot a_{0N}/a_0}{a_{0N}}).$$

$$(2.3.6)$$

Since

$$\frac{a_0}{a_{0N}} = \sqrt{\frac{\tau_0}{\tau_{0N}}}, \quad \text{and} \quad \frac{\rho_0 \tau_0}{\rho_{0N} \tau_{0N}} = \frac{p_0}{p_{0N}}, \tag{2.3.7}$$

the equation (2.3.6) yields:

$$a_D = c \cdot \frac{p_0}{p_{0N}} \cdot h_0(y) \cdot (3.927 \times 10^{-4} \rho_{0N}) \cdot (v \cdot \sqrt{\tau_{0N}/\tau_0})^2 C_D(\frac{v \cdot \sqrt{\tau_{0N}/\tau_0}}{a_{0N}}).$$

$$(2.3.8)$$

But,

$$G_D(v \cdot \sqrt{\tau_{0N}/\tau_0}) = (3.927 \times 10^{-4} \rho_{0N}) \cdot (v \cdot \sqrt{\tau_{0N}/\tau_0})^2 C_D(\frac{v \cdot \sqrt{\tau_{0N}/\tau_0}}{a_{0N}})$$

$$(2.3.9)$$

is the known G-function (2.3.3) obtained by formally replacing in it v with $v\sqrt{\tau_{0N}/\tau_0}$.

Thus, for the drag acceleration (2.3.8) of a given projectile in a non standard atmosphere we obtain:

$$a_D = c \cdot \frac{p_0}{p_{0N}} \cdot h_0(y) \cdot G_D(v \cdot \sqrt{\tau_{0N}/\tau_0}). \tag{2.3.10}$$

The vector differential equation (1.7.2), for a projectile flying in a non-standard atmosphere, is

$$\frac{d\vec{v}}{dt} = \vec{g} - c \cdot \frac{p_0}{p_{0N}} \cdot h_0(y) \cdot G_D(v \cdot \sqrt{\tau_{0N}/\tau_0}) \frac{\vec{v}}{v}, \tag{2.3.11}$$

where:

$$G_D(v\sqrt{\tau_{0N}/\tau_0}) = (3.927 \times 10^{-4}\rho_{0N})(v\cdot\sqrt{\tau_{0N}/\tau_0})^2 C_D(\frac{v\cdot\sqrt{\tau_{0N}/\tau_0}}{a_{0N}})$$

(2.3.12)

is the G-function of resistance obtained by formally replacing in (2.3.3) , the projectile speed v with ($v\sqrt{\tau_{0N}/\tau_0}$).

That means that the general form of the G-function is

$$G_D(v\cdot\sqrt{\tau_{0N}/\tau_0}) = \begin{cases} A\cdot(v\sqrt{\tau_{0N}/\tau_0})^2 & for \quad v\cdot\sqrt{\tau_{0N}/\tau_0} \le 256m/s \\ E\cdot(v\cdot\sqrt{\tau_{0N}/\tau_0} - F & for \quad v\cdot\sqrt{(\tau_{0N}/\tau_0)} > 256m/s \end{cases} .$$

(2.3.13)

For any atmosphere that is identical to ASM atmosphere, ICAO atmosphere, ASM atmosphere, or that is different from any of them, we can use the vector differential equation (2.3.11) to calculate the elements of the projectile trajectory.

Using the vector differential equation (2.3.11) we obtain the following systems of differential equations that describes the projectile trajectory in a non-standard atmosphere:

Variable t (time)

$$\begin{cases} \dfrac{dv_x}{dt} = -c\cdot\dfrac{p_0}{p_{0N}}\cdot h_0(y)\cdot G_D(v\cdot\sqrt{\tau_{0N}/\tau_0})\cdot\dfrac{v_x}{v} \\ \dfrac{dv_y}{dt} = -\dfrac{g}{v_x} \\ \dfrac{dx}{dt} = v_x \\ \dfrac{dy}{dt} = v_y \end{cases} ,$$

(2.3.14)

Variable x (abscissa)

$$\left\{ \begin{array}{l} \dfrac{dv_x}{dx} = -c \cdot \dfrac{p_0}{p_{0N}} h_0(y) \cdot \dfrac{G_D(v\sqrt{\tau_{0N}/\tau_0})}{v} \\[2ex] \dfrac{dp}{dx} = -\dfrac{g}{v_x^2} \\[2ex] \dfrac{dt}{dx} = \dfrac{1}{v_x} \\[2ex] \dfrac{dy}{dx} = p \end{array} \right. , (p = \tan\alpha) \quad (2.3.15)$$

Variable $p = \tan\alpha$ (α-angle of flight)

$$\left\{ \begin{array}{l} \dfrac{dv_x}{dp} = \dfrac{1}{g} \cdot c \dfrac{p_0}{p_{0N}} \cdot h_0(y) \cdot G_D(v\sqrt{\tau_{0N}/\tau_0}) \cdot \dfrac{v_x}{(1+p^2)^{1/2}} \\[2ex] \dfrac{dt}{dp} = -v_x/g \\[2ex] \dfrac{dx}{dp} = -v_x^2/g \\[2ex] \dfrac{dy}{dp} = -pv_x^2/g \end{array} \right. \quad . \quad (2.3.16)$$

where, at the sea level, or at the departing point of the projectile at a given altitude over the sea level:

- for the G-function that corresponds to the ASM atmosphere (section 1.4), the virtual temperature and the atmospheric pressure are respectively $\tau_{0N} = 289.60K$ (humidity 78%) and, $p_{0N} = 750\,\text{mmHg}$;
- for the G-function determined for the ICAO atmosphere (section 1.5), the virtual temperature and the atmospheric pressure are respectively $\tau_{0N} = 288.15K$ (humidity 0%) and, $p_{0N} = 760mmHg$;
- for the Siacci's G-function, Mayevski's G-function, or the G_{43}-function, the virtual temperature and the atmospheric pressure are respectively $\tau_{0N} = 289.08K$ (humidity 50%) and, $p_{0N} = 750$.

The density function is

$$h_0(y) = (\frac{\tau_0 - 0.006328y}{\tau_0})^{4.4}. \tag{2.3.17}$$

Note that the systems of differential equations (2.3.14), (2.3.15), and (2.3.16) can be used when the projectile is fired in any of the standard atmospheres, ASM, ICAO, or TSA atmosphere considering that at the firing site $p_0 = p_{0N}$ and $\tau_0 = \tau_{0N}$.

The origin of the rectangular system of coordinates is located at the muzzle of the firearm.

For $p_0 = p_{0N}$ and $\tau_0 = \tau_{0N}$ the differential equations (2.3.14), (2.3.15), and (2.3.16) yield respectively the systems of differential equations (2.1.3), (2.1.4), and (2.1.5).

Comments:

- The vector differential equation of the projectile flight (2.3.11) and the corresponding set of the scalar differential equations that can be obtained from that equation are already presented and used in the book *"Exterior Ballistics with Applications"* and in PC programs shown in the *"Exterior Ballistics of Small Arms"*.

- For a projectile flying in a standard atmosphere (ASM, ICAO, TSA) the differential equation (2.3.11) yields:

$$\frac{d\vec{v}}{dt} = \vec{g} - c \cdot h(y)G_D(v) \cdot \frac{\vec{v}}{v}. \tag{2.3.19}$$

2.4. Projectile-Streamline Model of the Trajectory

Hereafter we present an original approach to calculate the trajectory of a projectile fired in non standard atmosphere.

We assume that the characteristics of the atmosphere (virtual temperature, τ_0, air pressure, p_0, and air density, ρ_0) at the firing site, located not necessary at the sea level, are different from the respective characteristics of the standard atmosphere at the sea level (i.e., the virtual temperature, τ_{0N}, the pressure, p_{0N}, and the density, ρ_{0N}).

The density, as a function of the altitude "y" over the firing site, is

$$h_0(y) = (\frac{\tau_0 - 0.006328y}{\tau_0})^{4.4} .$$

The new method is based on two assumptions:

- On the model that considers the ballistics trajectory of flight as a streamline.
- In analogy with the streamline fluid flow, we assume that the flow field of the projectile trajectory-streamline can be superimposed to the gravitational field.

The validity of the projectile trajectory-streamline approach is demonstrated in examples of section 2.9.

The projectile-streamline approach can be applied to the mountain or high-altitude shootings, as well as to the winter low-temperature shootings, or to the summer high-temperature shootings.

The Model of the Projectile-Streamline

The acceleration of the projectile in flight in non-standard atmosphere with density, pressure, virtual temperature and sound speed respectively ρ_0, p_0, τ_0, and a_0, can be written:

$$\frac{d\vec{v}_2}{dt} = c \cdot h_0 \cdot 3.927 \times 10^{-4} \rho_0 v_2^2 C_D(\frac{v_2}{a_0}), \qquad (2.4.1)$$

where:

$$h_0(y) = (\frac{\tau_0 - 0.006328y}{\tau_0})^{4.4} .$$

The temperature changes with altitude "y" over the firing site according to the linear law:

$$\tau = \tau_0 - 0.006328 \cdot y .$$

At the firing site, a non-standard atmosphere is an atmosphere with characteristics different from the characteristics of the defined standard atmosphere, for example, a winter atmosphere, or a "mountain atmosphere".

Since we do not know the drag coefficient of the projectile in non-standard atmosphere, we need to express it through the known drag coefficient $C_D(v/a_{0N})$ of the projectile flying in a given standard atmosphere with characteristics ρ_{0N}, p_{0N}, τ_{0N}, and a_{0N} at the sea level altitude (firing site).

We model the projectile flight considering the trajectory as a fictive "flow" of identical point-mass projectiles launched continuously from the muzzle of firearm with the same initial speed, i.e., we assume that the projectile trajectory is a flow of point-mass projectiles.

We consider that:

• the trajectory of a point-mass projectile is a streamline of identical point-masses in motion (we use the term *"projectile trajectory-streamline"* to name the model we are using);
• the gravitational field is superimposed to the air flow field.

We can model the point-mass flow of a projectile using the Mach number (i.e., considering a model similar to the projectile "motion" in wind tunnels).

We assume that the flow field has kinematics similitude expressed by the equation:

$$M = \frac{v_1}{a_{0N}} = \frac{v_2}{a_0}. \tag{2.4.2}$$

Hence,

$$\frac{1}{a_{0N}}\frac{dv_1}{dt} = \frac{1}{a_0}\frac{dv_2}{dt}. \tag{2.4.3}$$

Using (2.4.2) and (2.4.3), for the drag acceleration (2.4.1), after a series of manipulations, we can write:

$$\frac{dv_1}{dt} = \frac{\rho_0}{\rho_{0N}} \frac{a_{0N}}{a_0} \cdot c \cdot h_0 \cdot (3.927 \times 10^{-4} \rho_{0N}) v_1^2 C_D(\frac{v_1}{a_{0N}}).$$

If we denote $v = v_1$, the above differential equation can be written:

$$\frac{dv}{dt} = \frac{\rho_0}{\rho_{0N}} \frac{a_{0N}}{a_0} \cdot c \cdot h_0 \cdot (3.927 \times 10^{-4} \rho_{0N}) v^2 C_D(\frac{v}{a_{0N}}). \qquad (2.4.4)$$

The G-drag function defined for the standard atmosphere is

$$G_D(v) = (3.927 \times 10^{-4} \rho_{0N}) \cdot v^2 \cdot C_D(\frac{v}{a_{0N}}). \qquad (2.4.5)$$

Substituting (2.4.5) in equation (2.4.4), we obtain:

$$\frac{dv}{dt} = \frac{\rho_0 \cdot a_{0N}}{\rho_{0N} \cdot a_0} \cdot c \cdot h_0 \cdot G_D(v). \qquad (2.4.6)$$

The differential equation (2.4.6) describes the projectile-streamline flow that has dynamic similitude with the projectile trajectory flying in the non-standard atmosphere.

The vector differential equation (2.1.1), i.e., the equation

$$\frac{d\vec{v}}{dt} = \vec{g} - c \cdot h(y) \cdot G_D(v) \frac{\vec{v}}{v}, \qquad (2.4.7)$$

will describe the projectile flight in non-standard atmosphere if we formally replace the ballistics coefficient c with the "fictive ballistics coefficient",

$$\bar{c} = \frac{\rho_0 \cdot a_{0N}}{\rho_{0N} \cdot a_0} \cdot c, \qquad (2.4.8)$$

and the density function h with h_0.

The differential equation (2.4.7) that describes the trajectory of the projectile flight in the given non-standard atmosphere can be written:

$$\frac{d\vec{v}}{dt} = \vec{g} - \bar{c} \cdot h_0(y) \cdot G_D(v)\frac{\vec{v}}{v} . \qquad (2.4.9)$$

The magnitude of the acceleration (2.4.6) is

$$\frac{dv}{dt} = \bar{c} \cdot h_0 \cdot G_D(v) . \qquad (2.4.10)$$

The fictive ballistics coefficient (2.4.8) can be written as:

$$\bar{c} = \frac{\rho_0}{\rho_{0N}} \sqrt{\frac{\tau_0}{\tau_{0N}}} \cdot c ,$$

or

$$\bar{c} = \frac{\rho_0}{\rho_{0N}} \sqrt{\frac{\tau_{0N}}{\tau_0}} \cdot c . \qquad (2.4.11)$$

The following system of differential equations that is obtained from the vector differential equation (2.4.7), describes the projectile flight in a non-standard atmosphere:

$$\begin{cases} \dfrac{dv_x}{dx} = -\bar{c} \cdot h_0(y) \cdot \dfrac{G_D(v)}{v} \\[2mm] \dfrac{dp}{dx} = -\dfrac{g}{v_x^2} \\[2mm] \dfrac{dt}{dx} = \dfrac{1}{v_x} \\[2mm] \dfrac{dy}{dx} = p \end{cases} , \qquad (2.4.12)$$

where

$$h_0(y) = (\frac{\tau_0 - 0.006328y}{\tau_0})^{4.4} , \qquad \bar{c} = \frac{\rho_0}{\rho_{0N}} \sqrt{\frac{\tau_{0N}}{\tau_0}} \cdot c . \qquad (2.4.13)$$

In general, the solution of the differential equations of projectile flight is obtained using the numerical integration executed by a PC program.

To illustrate the theoretical results and to solve exterior ballistics problems, at the end of the section are presented the PC programs ACA122.BAS, and RAN122.BAS.

The PC programs, ACA122.BAS and RAN122.BAS, are modified versions respectively of the PC programs Angmet122.Bas and Rangmet.Bas. [14].

The programs can be used to solve the differential equations (2.4.12) only for a Russian 122mm (HE) projectile fired by a cannon Mod.1960 with initial speed 885m/s (main propellant charge).

The solutions of the differential equations of the projectile flight obtained using the above programs are identical to the solutions obtained using Angmet122.Bas and Rangmet.Bas.

The PC programs show as well the ballistics coefficient and the fictive ballistics coefficient as a function of the departure angle.

Both programs can be modified to be used with any firearm when the ballistics coefficient is known as a function of departure angle, or as a function of the projectile speed.

Example 4.1 Fictive Ballistics Coefficient

For a 7.62mm-M80 Ball, 148-grain FMJBT bullet for a 24"'" Barrel in ASM atmosphere:

temperature, 59 degrees Fahrenheit (15 degrees Celsius); pressure, 29.53 inches (750mmHg); humidity, 78%; (pressure of water vapors, 9.906mm; virtual temperature, (289.6 degree Celsius), the ballistics coefficient (relative to G_1-function) is 0.404 lb/in2. The departure speed of the projectile is 2854 ft/s (870m/s).

---

14. Klimi, G., "Exterior Ballistics of Small Arms", p. 151, p.235, Xlibris 2009

Find the fictive ballistics coefficient that corresponds to the ICAO standard atmosphere if:

- at the firing site, there is a hot and a very humid summer time weather: temperature of dry air, 90 degrees Fahrenheit (32.2 degrees Celsius); pressure 29.724 in. (755 mm Hg), relative humidity, 90%;
- the firing site is located in a mountain at an altitude of 3950 feet (1204m), and there is winter weather: pressure is 25.4 in. Hg (646 mmHg), temperature is -15 degrees Fahrenheit (-26 degrees Celsius); humidity is 30%.

Solution:

The ballistics coefficient in metric units is

$$c = \frac{1.4222}{C} = \frac{1.4222}{0.404} = 3.5203 m^2 \, / \, kg.$$

ICAO Atmosphere

Pressure, $p_0 = 760 mm \ Hg.$, relative humidity, 0% ,; speed of sound, $a_0 = 340.30 m \, / \, s$,; temperature, $t_0 = 15° \ C$,; air density, $\rho_0 = 1.225 kg \, / \, m^3$.

$$\bar{c} = \frac{p_0}{p_{0N}} \sqrt{\frac{\tau_{0N}}{\tau_0}} \cdot c = \frac{760}{750} \sqrt{\frac{289.6}{288.15}} \cdot (3.5203) = 3.5762 \, .$$

In English units

$$\bar{C} = \frac{1.4222}{\bar{c}} = \frac{1.4222}{3.5762} = 0.3977 \, .$$

High-temperature summer shooting

Using table 1, section 2.2, and the formula (2.2.1), we find that the pressure of water vapors is 30.50 mm Hg, while the virtual temperature is

$$\tau_0 = \frac{T_0}{(1 - 0.3785 \cdot e_0 / p_0)} = \frac{273.15 + 32.22}{(1 - 0.3785 \cdot (30.50) / 755)} = 310.12^\circ \text{ Kelvin.}$$

The fictive ballistics coefficient is

$$\bar{c} = \frac{p_0}{p_{0N}} \sqrt{\frac{\tau_{0N}}{\tau_0}} \cdot c = \frac{755}{750} \sqrt{\frac{289.6}{310.12}} \cdot (3.5203) = 3.4245 .$$

The fictive BC in English units

$$\bar{C} = \frac{1.4222}{\bar{c}} \frac{1.4222}{3.4245} = 0.4152 .$$

High-mountain winter shooting

In the same way as above, we find that the fictive BC is

$$\bar{c} = \frac{p_0}{p_{0N}} \sqrt{\frac{\tau_{0N}}{\tau_0}} \cdot c = \frac{646}{750} \sqrt{\frac{289.6}{258}} \cdot (3.5203) = 3.1241 .$$

The fictive BC in English units is

$$\bar{C} = \frac{1.4222}{\bar{c}} \frac{1.4222}{3.1241} = 0.4552 .$$

PC QuickBasic Program
ACA122.BAS

```
'FIND:  Departure Angle, Time of Flight, y-coordinate of a point, etc.
'GIVEN: Coordinates of target and the Cannon, Ballistics Coefficient as function
of departure angle
'Departure Speed
'Range-Wind, Cross-Wind are Present
'-------------------------------------------------------------------------------------
'CONTROL DATA
'INPUT:   x-coordinate of TARGET [m] = 14,200; y-coord of cannon = 0;
'Departure Speed = 885; x-coordinate of a point on Trajectory = 5500
'Range-Wind = 0; Cross-Wind = 0; Temperature of Air = 6;
'Propellant Temperature = 15; Atmospheric Pressure = 626;
'Pressure of water Vapors = 3.4; Projectile Standard Mass = 27.3;
'Change in Projectile Mass = 0;


'RESULTS: Launching Angle = 10.28125 [Degree], Time of Flight = 26.34 [Sec.]
'Terminal speed = 368; Terminal Angle = -16.3318 [Degree]
'Coordinates of Trajectory vertex (8171, 865)
'For x = 5500m: y = 751m; Time = 7.48s; Angle = 4.608 Degree;
'Speed = 625m/s
'-------------------------------------------------------------------------------------
' Note: Round the Input RANGE to the nearest 1;
'-------------------------------------------------------------------------------------
'Functions and Sub. Prog.

DECLARE SUB y1z1v1w1 (x, y, z, v, w, y1, z1, v1, w1, koef, yy, pa1)
DECLARE SUB InfHyres (xx, n, koef, vv, vo, yo, xc1, cw, wind, ta, tp, pa, ea,
m, dm, cf, voo, pa1)
DECLARE SUB InfDales (x0, y0, z0, v0, w0, xm, ym, A, xc, yc, tc, vc, ac, vo,
vv, xx, vo, cw, BC, BC1)
DECLARE SUB NPxyzvw (nk, x, x0, y, y0, z, z0, v, v0, w, w0, h, h0, k, l, r, q)
DECLARE SUB NPkoef (k, l, r, q, h, y1, z1, v1, w1)
DECLARE SUB menu (cog, cof, xf, yf, xfu, yfu, t$)
DECLARE SUB y0z0 (y0, z0, A, vo, wind)
DECLARE SUB C (koef, A, cf, dm, m, BC, BC1)

'Variables
SCREEN 0
1 :
DIM m(4, 4), v(4)          'Intermediate values (k,l,r,q)
rendi = 4                  'rend dif.
cog = 7: cof = 0
cikli = 0
A = 23                     'Initial Angle 23 degree
```

```
kendi = 22              'd.Angle for maximum distance
kov = 1                 'Test of the value of v0
gab = 1                 'error 1m.
tt = 1
'Solution
CLS
'Initial Data
menu cog, cof, 3, 10, 7, 70, "DATA INPUT"
InfHyres xx, n, koef, vv, vo, yo, xc1, cw, wind, ta, tp, pa, ea, m, dm, cf, voo, pa1
hap = 1
'Initial values
f:
x0 = 0: v0 = vv: w0 = 0
y0z0 y0, z0, A, vo, wind: h0 = hap
C koef, A, cf, dm, m, BC, BC1
ff:
FOR nk = 1 TO rendi
NPxyzvw nk, x, x0, y, y0, z, z0, v, v0, w, w0, h, h0, k, l, r, q
y1z1v1w1 x, y, z, v, w, y1, z1, v1, w1, koef, yy, pa1
NPkoef k, l, r, q, h, y1, z1, v1, w1
m(nk, 1) = k: m(nk, 2) = l
m(nk, 3) = r: m(nk, 4) = q
NEXT nk
'Estimations for new points
FOR i = 1 TO rendi
v(i) = 1 / 6 * (m(1, i) + 2 * m(2, i) + 2 * m(3, i) + m(4, i))
NEXT i

'New Points
x0 = x0 + h: y0 = y0 + v(1): z0 = z0 + v(2)
v0 = v0 + v(3): w0 = w0 + v(4)
xcc = x0 + wind * w0
IF ABS(xcc - xc1) <= .5 THEN
xc = xcc
yc = v0
tc = w0
ac = (180 / 3.141592654#) * ATN(z0)
vc = y0 / COS(ATN(z0))
END IF
xmm = x0 + w0 * wind
IF xmm > 30 AND ABS(z0) <= .00001 THEN
xm = xmm
ym = v0
END IF

'Tests the y-value
xT = x0 + wind * w0
```

```
IF kov = 1 THEN kov = -1: GOTO ff:
IF ABS(xT - xx) < gab AND v0 <= yo + (gab * TAN(A * 3.1415954# / 180))
AND v0 >= yo + ((-1 * gab) * TAN(A * 3.1415954# / 180)) THEN
C:
'Display Results
CLS
PLAY "a8a16a32b8"
menu 12, 0, 4, 10, 11, 70, "RESULTS:"
COLOR 7
InfDales x0, y0, z0, v0, w0, xm, ym, A, xc, yc, tc, vc, ac, yo, vv, xx, vo, cw, BC,
BC1
CLS
GOTO 1:
END IF

IF ABS(xT - xx) < gab AND v0 > yo + (gab * TAN(A * 3.1415954# / 180))
THEN
t$ = "  * ? *"
menu 18, 0, 10, 20, 14, 60, t$
COLOR 14
LOCATE 12, 30: PRINT "Wait a moment, Please (+)";
LOCATE 12, 53: PRINT tt
tt = tt + 1
COLOR 7
A = A - kendi
GOTO fff:
END IF

IF ABS(xT - xx) < gab AND v0 < yo + ((-1 * gab) * TAN(A * 3.1415954# /
180)) THEN
t$ = "  * ? *"
menu 18, 0, 10, 20, 14, 60, t$
COLOR 14
LOCATE 12, 30: PRINT "Wait a moment, Please (-)";
LOCATE 12, 53: PRINT tt
tt = tt + 1
COLOR 7
A = A + kendi
GOTO fff:
END IF
GOTO ff:
fff:
'Restart Cycle
cikli = cikli + 1
IF cikli = 20 THEN GOTO C:
kendi = kendi / 2
kov = 1
```

GOTO f:

```
SUB C (koef, A, cf, dm, m, BC, BC1)
IF A >= 0 AND A <= 1.38333333# THEN koef = (1 - dm / m) * (1.82051 -
210.432 * (A * 3.141592654# / 180) + 10066.5 * (A * 3.141592654# / 180) ^ 2 -
164950 * (A * 3.141592654# / 180) ^ 3)
IF A > 1.38333333# AND A <= 2.5666667# THEN koef = (1 - dm / m) *
(.490432 - 14.0538 * (A * 3.141592654# / 180) + 298.238 * (A * 3.141592654# /
180) ^ 2 - 2328.38 * (A * 3.141592654# / 180) ^ 3)
IF A > 2.5666667# AND A <= 4.8833333# THEN koef = (1 - dm / m) *
(.352739 - 4.0664 * (A * 3.141592654# / 180) + 49.8947 * (A * 3.141592654# /
180) ^ 2 - 207.518 * (A * 3.141592654# / 180) ^ 3)
IF A > 4.8833333# AND A < 8.0066667# THEN koef = (1 - dm / m) * (.304165
- 1.65929 * (A * 3.141592654# / 180) + 13.5957 * (A * 3.141592654# / 180) ^ 2
- 35.3659 * (A * 3.141592654# / 180) ^ 3)
IF A > 8.0066667# AND A <= 12.9166667# THEN koef = (1 - dm / m) *
(.324563 - 1.5842 * (A * 3.141592654# / 180) + 9.56496 * (A * 3.141592654# /
180) ^ 2 - 17.802 * (A * 3.141592654# / 180) ^ 3)
IF A > 12.9166667# AND A <= 18.3166667# THEN koef = (1 - dm / m) *
(.158193 + .919854 * (A * 3.141592654# / 180) - 3.03889 * (A * 3.141592654# /
180) ^ 2 + 3.35924 * (A * 3.141592654# / 180) ^ 3)
IF A > 18.3166667# AND A <= 45 THEN koef = (1 - dm / m) * (.251064 -
.0064118# * (A * 3.141592654# / 180) + .0262451# * (A * 3.141592654# / 180)
^ 2 - .0064443# * (A * 3.141592654# / 180) ^ 3)
BC = koef
koef = cf * koef
BC1 = koef
END SUB


SUB InfDales (x0, y0, z0, v0, w0, xm, ym, A, xc, yc, tc, vc, ac, yo, vv, xx, vo,
cw, BC, BC1)
aT = ATN(z0) * 180 / 3.141592654#
LOCATE 5, 18: PRINT "Departure Angle    = "; A; " degree"
LOCATE 6, 18: PRINT "Time of Flight     = "; INT((w0) * 1000 + .5) / 1000; "
seconds";
LOCATE 7, 18: PRINT "Terminal Speed     = "; INT((y0 / COS(ATN(z0))) *
100 + .5) / 100; "m/s"
LOCATE 8, 18: PRINT "Terminal Angle     = "; aT; " degree"
LOCATE 9, 18: PRINT "Trajectory Vertex  = :"; "("; INT((xm) * 100 + .5) / 100;
","; INT((ym) * 100 + .5) / 100; ")"
LOCATE 10, 18: PRINT "Cross-Wind Deflection     = "; INT((cw * (w0 - xx /
(vo * COS(A * 3.14159265# / 180)))) * 1000 + .5) / 1000
LOCATE 11, 18: PRINT "The Ballistics Coefficient :"; BC
LOCATE 13, 16: PRINT "Distance to the Target    = "; INT(((xx ^ 2 + (vv - v0)
^ 2) ^ .5) * 100 + .05) / 100; " meter";
LOCATE 14, 16: PRINT "x, y coordinates of TARGET = "; "("; INT((xx) * 100
+ .5) / 100; ","; INT((v0) * 100 + .5) / 100; ")"
```

LOCATE 15, 16: PRINT "x,y coordinates of GUN   = "; "("; 0; ","; vv; ")"
LOCATE 17, 18: PRINT "x-coordinate of a Point[m] :"; INT((xc) * 100 + .5) / 100
LOCATE 18, 18: PRINT "Corresponding y [m] :"; INT((yc) * 1000 + .5) / 1000
LOCATE 19, 18: PRINT "Corresponding Time [sec] :"; INT((tc) * 100 + .5) / 100
LOCATE 20, 18: PRINT "Corresponding Speed [m/s] :"; INT((vc) * 100 + .5) / 100
LOCATE 21, 18: PRINT "Corresponding Angle [Deg]   :"; ac
LOCATE 22, 18: PRINT "Cross-Wind Deflection   :"; INT((cw * (tc - xc / (vo
* COS(A * 3.14159265# / 180)))) * 1000 + .5) / 1000
LOCATE 23, 18: PRINT "The Fictive BC :"; BC1
COLOR 7
LOCATE 24, 11: PRINT " Pres [P] to repeat [Esc] to end ";
cc$ = INPUT$(1)
IF cc$ = CHR$(27) THEN SCREEN 9: CLS : END
END SUB

SUB InfHyres (xx, n, koef, vv, vo, yo, xc1, cw, wind, ta, tp, pa, ea, m, dm, cf,
voo, pa1)
LOCATE 4, 13: INPUT "x-coordinate of TARGET [m]   ="; xx
LOCATE 5, 13: INPUT "y-coordinate of TARGET [m]   ="; yo
LOCATE 6, 13: INPUT "y-coord of Cannon   ="; vv
LOCATE 7, 13: INPUT "Departure Speed [m/s]   ="; voo
LOCATE 9, 13: INPUT "X-coordinate of a Point   ="; xc1
LOCATE 10, 13: INPUT "Temperature of Dry Air [C] = "; ta
LOCATE 11, 13: INPUT "Propellant Temperature [C] = "; tp
LOCATE 12, 13: INPUT "Atmospheric Pressure [mmHg] = "; pa
LOCATE 13, 13: INPUT "Pressure of water Vapors   = "; ea
LOCATE 14, 13: INPUT "Projectile Standard Mass[kg] = "; m
LOCATE 15, 13: INPUT "Change in Projectile Mass   = "; dm
LOCATE 16, 13: INPUT "Range-Wind   = "; wind
LOCATE 17, 13: INPUT "Cross-Wind   = "; cw
ta = ta + 273.15
pa1 = ta / (1 - .3785 * ea / pa)
cf = (pa / 750) * (289.08 / pa1) ^ .5
vo = (voo - .4 * voo * (dm / m) + .001285 * voo * (tp - 15))
CLS
END SUB

SUB menu (cog, cof, xf, yf, xfu, yfu, t$)
COLOR cog, cof
LOCATE xf - 1, yf: PRINT t$
LOCATE xf, yf: PRINT "É" + STRING$(yfu - yf, 205) + "»";
FOR i = xf + 1 TO xfu
LOCATE i, yf: PRINT "º" + SPACE$(yfu - yf) + "º";
NEXT
LOCATE xfu + 1, yf: PRINT "È" + STRING$(yfu - yf, 205) + "¼";
END SUB

```
SUB NPkoef (k, l, r, q, h, y1, z1, v1, w1)
k = h * y1: l = h * z1
r = h * v1: q = h * w1
END SUB

SUB NPxyzvw (nk, x, x0, y, y0, z, z0, v, v0, w, w0, h, h0, k, l, r, q)
IF nk = 1 THEN
x = x0: y = y0: z = z0
v = v0: w = w0: h = h0
GOTO fund:
END IF
IF nk = 2 OR nk = 3 THEN
x = x0 + (.5 * h): y = y0 + (.5 * k)
z = z0 + (.5 * l): v = v0 + (.5 * r)
w = w0 + (.5 * q)
GOTO fund:
END IF
IF nk = 4 THEN
x = x0 + h: y = y0 + k: z = z0 + l
v = v0 + r: w = w0 + q
END IF
fund:
END SUB

SUB y0z0 (y0, z0, A, vo, wind)
y0 = SQR(vo ^ 2 + wind ^ 2 - 2 * vo * wind * COS(A * 3.141592654# / 180))
y0 = y0 * COS(A * 3.141592654# / 180)
z0 = TAN(A * 3.141592654# / 180)
z0 = z0 / (1 - wind / (vo * COS(A * 3.141592654# / 180)))
END SUB
SUB y1z1v1w1 (x, y, z, v, w, y1, z1, v1, w1, koef, yy, pa1)
yy = y * SQR(1 + z ^ 2)
IF yy > 256! THEN
y1 = -1 * koef * ((pa1 - .006328 * v) / pa1) ^ 4.4 * (yy - 240) / (3 * yy)
ELSE
y1 = -1 * koef * ((pa1 - .006328 * v) / pa1) ^ 4.4 * .0001212 * yy ^ 2 / yy
END IF
z1 = -9.80665 / y ^ 2
v1 = z
w1 = 1 / y
END SUB
```

QBasic PC Program
RAN122.BAS

```
'FIND : Range and other Elements of the Trajectory, etc.
'GIVEN: Departure Speed, Departure Angle, Ballistics Coefficient
'The PC program RAN122.BAS can be used to solve the exterior ballistics
problems
'in any atmospheric conditions.
'RAN122.BAS considers the pressure, temperature of air at the firing site (at or
above the sea
 level)
'The Cartesian Coordinate system has the origin at the muzzle of the firearm.
'The y-coordinate of the firearm is always equal to zero.
'------------------------------------------------------------------------------------
' DATA
' Input:  Initial y-coordinate = 0, departure Angle; 5.6205, departure speed ,885
'Temperature of Air, 6, temperature of propellant, 15;
'Pressure = 626, Pressure of Air vapor = 3.4, Projectile mass, 23;
'Change in Projectile mass = 0. Wind =0
'
'Results: Range = 10000, Error in y-coordinate,-0.06, Time of Flight = 15.66,
'Terminal Speed = 472, Terminal Angle = -8.5419 Degree
'Cross wind deflection, 0; vertex (5532,302)
'Ballistics Coefficient, 0.2388; Fictive Ballistics Coefficient, 0.2027
'------------------------------------------------------------------------------------
'Functions & Subs.
DECLARE SUB y1z1v1w1 (x, y, z, v, w, y1, z1, v1, w1, koef, yy, pa1)
DECLARE SUB InfHyres (x0, y0, z0, v0, w0, A, h0, ta, pa, ea, m, dm, tp, ta1,
pa1, xx1, voo, vo1, wind, cf, cw, vv)
DECLARE SUB NPxyzvw (nk, x, x0, y, y0, z, z0, v, v0, w, w0, h, h0, k, L, r, q)
DECLARE SUB NPkoef (k, L, r, q, h, y1, z1, v1, w1)
DECLARE SUB menu (cog, cof, xf, yf, xfu, yfu, t$)
DECLARE SUB c (koef, m, dm, BC, A, cf, BC1)

'Variables
DIM m(4, 4), v(4)
rendi = 4
cog = 7: cof = 0

'Zgjidhja
CLS
fillimi:
menu cog, cof, 3, 10, 21, 70, "INITIAL DATA"
InfHyres x0, y0, z0, v0, w0, A, h0, ta, pa, ea, m, dm, tp, ta1, pa1, xx1, voo, vo1,
wind, cf, cw, vv
c koef, m, dm, BC, A, cf, BC1
```

```
f:
FOR nk = 1 TO rendi
NPxyzvw nk, x, x0, y, y0, z, z0, v, v0, w, w0, h, h0, k, L, r, q
y1z1v1w1 x, y, z, v, w, y1, z1, v1, w1, koef, yy, pa1
NPkoef k, L, r, q, h, y1, z1, v1, w1
m(nk, 1) = k: m(nk, 2) = L
m(nk, 3) = r: m(nk, 4) = q
NEXT nk

'Calculation
FOR i = 1 TO rendi
v(i) = 1 / 6 * (m(1, i) + 2 * m(2, i) + 2 * m(3, i) + m(4, i))
NEXT i

'New Data
x0 = x0 + h: y0 = y0 + v(1): z0 = z0 + v(2)
v0 = v0 + v(3): w0 = w0 + v(4)

IF ABS(z0) < .0001 THEN
ymax = v0
xmax = x0 + wind * w0
END IF


xxc = x0 + wind * w0
IF (xxc - xx1) <= .001 THEN
xc = xxc
yc = v0
tc = w0
ac = (180 / 3.141592654#) * ATN(z0)
vc = y0 / COS(ATN(z0))
END IF


IF v0 - vv <= .01 THEN
'Display Results
menu cog, cof, 6, 20, 22, 72, "RESULTS:"
LOCATE 7, 25: PRINT "Horizontal Range [m]    = "; INT((x0 + w0 * wind) + .5)
LOCATE 8, 25: PRINT "Coresponding y-Coord [m] = "; INT((v0) * 1000 + .5) /
1000
LOCATE 9, 25: PRINT "Departure Angle [Deg.]   = "; INT((A) * 1000000 + .5)
/ 1000000
LOCATE 10, 25: PRINT "Time of Flight [s]      = "; INT((w0) * 100 + .5) / 100
LOCATE 11, 25: PRINT "Terminal Speed [m/s]    = "; INT((y0 * (1 + z0 ^ 2) ^
.5) + .5)
LOCATE 12, 25: PRINT "Terminal Angle [Deg.]   = "; INT((ATN(z0) * 180 /
3.141593) * 10000 + .5) / 10000
LOCATE 13, 25: PRINT "Cross-Wind Deflection   = "; INT((cw * (w0 - x0 /
(voo * COS(A * 3.14159265# / 180)))) * 1000 + .5) / 1000
```

LOCATE 14, 25: PRINT "Trajectory Vertex [m]   = "; "("; INT((xmax) + .5); ","; INT((ymax) + .5); ")"
LOCATE 16, 25: PRINT "Point on Trajectory [m]   = "; "("; INT((xc) + .5); ","; INT((yc) * 1000 + .5) / 1000; ")"
LOCATE 17, 25: PRINT "Time [s]   = "; INT((tc) * 100 + .5) / 100
LOCATE 18, 25: PRINT "Corresponding Speed [m/s] = "; INT((vc) + .5)
LOCATE 19, 25: PRINT "Corresponding Angle [Deg] = "; INT((ac) * 10000 + .5) / 10000
LOCATE 20, 25: PRINT "Cross-Wind Deflection   = "; INT((cw * (tc - xc / (voo * COS(A * 3.14159265# / 180)))) * 1000 + .5) / 1000
LOCATE 22, 25: PRINT "Ballistics Coefficient   = "; BCLOCATE 23, 25: PRINT "Fictive BC   = "; BC1
ELSE
GOTO f:
END IF
END

SUB c (koef, m, dm, BC, A, cf, BC1)
IF A >= 0 AND A <= 1.38333333# THEN koef = (1 - dm / m) * (1.82051 - 210.432 * (A * 3.141592654# / 180) + 10066.5 * (A * 3.141592654# / 180) ^ 2 - 164950 * (A * 3.141592654# / 180) ^ 3)
IF A > 1.38333333# AND A <= 2.5666667# THEN koef = (1 - dm / m) * (.490432 - 14.0538 * (A * 3.141592654# / 180) + 298.238 * (A * 3.141592654# / 180) ^ 2 - 2328.38 * (A * 3.141592654# / 180) ^ 3)
IF A > 2.5666667# AND A <= 4.8833333# THEN koef = (1 - dm / m) * (.352739 - 4.0664 * (A * 3.141592654# / 180) + 49.8947 * (A * 3.141592654# / 180) ^ 2 - 207.518 * (A * 3.141592654# / 180) ^ 3)
IF A > 4.8833333# AND A < 8.0066667# THEN koef = (1 - dm / m) * (.304165 - 1.65929 * (A * 3.141592654# / 180) + 13.5957 * (A * 3.141592654# / 180) ^ 2 - 35.3659 * (A * 3.141592654# / 180) ^ 3)
IF A > 8.0066667# AND A <= 12.9166667# THEN koef = (1 - dm / m) * (.324563 - 1.5842 * (A * 3.141592654# / 180) + 9.56496 * (A * 3.141592654# / 180) ^ 2 - 17.802 * (A * 3.141592654# / 180) ^ 3)
IF A > 12.9166667# AND A <= 18.3166667# THEN koef = (1 - dm / m) * (.158193 + .919854 * (A * 3.141592654# / 180) - 3.03889 * (A * 3.141592654# / 180) ^ 2 + 3.35924 * (A * 3.141592654# / 180) ^ 3)
IF A > 18.3166667# AND A <= 45 THEN koef = (1 - dm / m) * (.251064 - .0064118# * (A * 3.141592654# / 180) + .0262451# * (A * 3.141592654# / 180) ^ 2 - .0064443# * (A * 3.141592654# / 180) ^ 3)
BC = koef / (1 - dm / m)
koef = cf * koef
BC1 = koef
END SUB

SUB InfHyres (x0, y0, z0, v0, w0, A, h0, ta, pa, ea, m, dm, tp, ta1, pa1, xx1, voo, vo1, wind, cf, cw, vv)
LOCATE 5, 13: INPUT "y-coordinate of Firearm = "; v0

```
LOCATE 6, 13: INPUT "Departure Speed [m/s]    = "; y0
LOCATE 7, 13: INPUT "Departure Angle [Degree] = "; z0
LOCATE 8, 13: INPUT "Temperature of Air [C]      = "; ta
LOCATE 9, 13: INPUT "Propellant Temperature[C]    = "; tp
LOCATE 10, 13: INPUT "Atmospheric Pressure [mm]   = "; pa
LOCATE 11, 13: INPUT "Pressure of Water Vapor [mm] = "; ea
LOCATE 12, 13: INPUT "Projectile Mass          = "; m
LOCATE 13, 13: INPUT "Change in Projectile mass   = "; dm
LOCATE 14, 13: INPUT "Range Wind               = "; wind
LOCATE 15, 13: INPUT "Cross Wind               = "; cw
LOCATE 16, 13: INPUT "x-coordinate of a point on Trajectory = "; xx1
LOCATE 17, 13: INPUT "Integration Step,  10, 1, or 0.5 = "; h0
vv = v0: A = z0: voo = y0
ta = ta + 273.15
pa1 = ta / (1 - .3785 * ea / pa)
cf = (pa / 750) * (289.08 / pa1) ^ .5
vo = (voo - .4 * voo * (dm / m) + .001285 * voo * (tp - 15))
vo1 = (voo - .4 * voo * (dm / m) + .001285 * voo * (tp - 15))
y0 = SQR(vo1 ^ 2 + wind ^ 2 - 2 * vo1 * wind * COS(A * 3.141592654# / 180))
y0 = y0 * COS(A * 3.141592654# / 180)
z0 = TAN(A * 3.141592654# / 180)
z0 = z0 / (1 - wind / (vo1 * COS(A * 3.141592654# / 180)))
CLS
END SUB


SUB menu (cog, cof, xf, yf, xfu, yfu, t$)
COLOR cog, cof
LOCATE xf - 1, yf: PRINT t$
LOCATE xf, yf: PRINT "É" + STRING$(yfu - yf, 205) + "»";
FOR i = xf + 1 TO xfu
LOCATE i, yf: PRINT "º" + SPACE$(yfu - yf) + "º";
NEXT
LOCATE xfu + 1, yf: PRINT "È" + STRING$(yfu - yf, 205) + "¼";
END SUB


SUB NPkoef (k, L, r, q, h, y1, z1, v1, w1)
k = h * y1: L = h * z1
r = h * v1: q = h * w1
END SUB


SUB NPxyzvw (nk, x, x0, y, y0, z, z0, v, v0, w, w0, h, h0, k, L, r, q)
IF nk = 1 THEN
x = x0: y = y0: z = z0
v = v0: w = w0: h = h0
GOTO fund:
END IF
IF nk = 2 OR nk = 3 THEN
```

```
x = x0 + (.5 * h): y = y0 + (.5 * k)
z = z0 + (.5 * L): v = v0 + (.5 * r)
w = w0 + (.5 * q)
GOTO fund:
END IF
IF nk = 4 THEN
x = x0 + h: y = y0 + k: z = z0 + L
v = v0 + r: w = w0 + q
END IF
fund:
END SUB


SUB y1z1v1w1 (x, y, z, v, w, y1, z1, v1, w1, koef, yy, pa1)
yy = y * SQR(1 + z ^ 2)
IF yy > 256! THEN
y1 = -1 * koef * ((pa1 - .006328 * v) / pa1) ^ 4.4 * (yy - 240) / (3 * yy)
ELSE
y1 = -1 * koef * ((pa1 - .006328 * v) / pa1) ^ 4.4 * .0001212 * yy ^ 2 / yy
END IF
z1 = -9.80665 / y ^ 2
v1 = z
w1 = 1 / y
END SUB
```

2.5. Conversion of Ballistics Coefficient to Standard Value

The fictive ballistics coefficient $\bar{c}$, given in (2.4.11), depends on the temperature and pressure at the firing site and their respective standard values at the sea level.

The value of the fictive ballistics coefficient measured experimentally (for a given departure angle and departure speed) in non-standard atmosphere shows a "deviation" in value from the respective value of the ballistics coefficient c measured at the sea level in standard atmosphere.

The small arms shooters or ballisticians that experiment and "calculate their own" BC of bullets in non-standard atmospheric conditions (for example,: in low temperatures of winter, high temperatures of summer, high-mountain shooting, etc.) find experimentally a ballistics coefficient that in general is "different" from the one that the manufacturer claims. [15].

As a matter of fact, at the firing site, the small arms shooters measure the "fictive ballistics coefficient" $\bar{c}$ of the projectile,

$$\bar{c} = \frac{p_0}{p_{0N}} \sqrt{\frac{\tau_{0N}}{\tau_0}} \cdot c, \qquad (2.5.1)$$

that (for a given departure angle) is a function of temperature and pressure (or equivalently, it is a function of density of air and speed of sound).

Recall that based on the definition of the ballistics coefficient,

$$c = \frac{i \cdot d^2}{m} 1000 ,$$

15.   - Hayden, R, Almgren , T Thomas, K, McDonald W. T. "Exterior Ballistics Explained", Lessons Learned from Ballistic Coefficient Testing, http://www.exteriorballistics.com/ebexplained/5th/24.cfm (Web accessed July 2009
- Stephen Ricciardelly, Steves page: http://stevespages.com/bc.html (web access July, 2009)

we can state that:

- the ballistics coefficient c that corresponds to a given G-function is a ballistics characteristic of the given standard projectile;
- ballistics coefficient c does not change as result of the deviation of the atmospheric characteristics from the characteristics of the referred standard atmosphere.

In other words, the ballistics coefficient c of a projectile (related with a given G-function) remains the same no matter what are the characteristics of atmosphere.

It is the fictive ballistics coefficient $\bar{c}$ that changes with the variation of the characteristics of atmosphere.

Thus, to find the value c of the ballistics coefficient we "convert" the experimentally measured fictive ballistics coefficient $\bar{c}$ into the ballistics coefficient c, solving the equation (2.5.1).

Solving (2.5.1), we obtain the ballistics coefficient

$$c = \frac{p_{0N}}{p_0} \sqrt{\frac{\tau_0}{\tau_{0N}}} \cdot \bar{c} \qquad (2.5.2)$$

that corresponds to the sea level shooting in standard atmosphere.

Actually, using (2.5.2), we do not "convert" $\bar{c}$ into c, but we find the value c of the ballistics coefficient.

Note that the fictive BC in English units is

$$\bar{C} = \frac{p_{0N}}{p_0} \sqrt{\frac{\tau_0}{\tau_{0N}}} \cdot C, \qquad (2.5.3)$$

The standard atmosphere BC of the projectile in English units is related to the respective fictive BC by the formula

$$C = \frac{p_0}{p_{0N}} \sqrt{\frac{\tau_{0N}}{\tau_0}} \cdot \overline{C} . \qquad (2.5.4)$$

It is important to note that using (2.5.2) or (2.5.4), we can find the BC of a bullet, or, in general, the BC of a projectile using the experimental results obtained in whatever atmospheric conditions, not necessary in standard conditions or in a special proving- ground laboratory.

Example 5.1 Conversion of the BC of a Bullet

The fictive ballistics coefficient of a 180-grain, caliber 0.30"-caliber" bullet, measured as result of firing tests at the altitude 5,000 feet over the sea level, is $\overline{C}$=0.4982lb/in^2 (the BC corresponds to the G_1-function, ASM atmosphere).

The departure speed of the projectile at the sea level is 2550ft/s. The temperature at the firing site is -20 degrees Fahrenheit, and the humidity is 40%. Pressure at the shooting site is 24.49 inches of Hg.

Find the ballistics coefficient of the bullet in ASM standard atmosphere (pressure 29.53 inches of Hg,; temperature 59 degrees Fahrenheit, humidity 78%; the virtual temperature at the sea level is 290 degrees Kelvin), converting the measured BC into the sea level altitude.

Solution:

The temperature in degree Celsius is

$$T_C = \frac{5}{9}(F - 32) = \frac{5}{9}(-20 - 32) = -28.89° .$$

The temperature in Kelvin is

$$T_C = 273.15° + (-28.89°) = 244.26° \, K$$

Using table 1, section 2.3, we find that the pressure of water vapors is e_0=0.197$mmHg$. The virtual temperature is

$$\tau = \frac{T}{1 - 0.3785(e/p)} = \frac{244.26}{1 - 0.3785(0.197/622)} = 244.29 \; ^\circ Kelvin.$$

We find that the fictive ballistics coefficient is

$$C = \frac{p_0}{p_{0N}}\sqrt{\frac{\tau_{0N}}{\tau_0}} \cdot \overline{C} = \cdot \frac{24.49}{29.53}\sqrt{\frac{290}{244.29}} = 0.9036(0.4982) = 0.4502 lb/in^2.$$

2.6. Projectile Trajectory in Mountains, in High or Low Temperatures

Mountain Firing

Assuming that the projectile is launched at a site in altitude "y_a" over the sea level, with the characteristics of atmospheric air ρ_0, p_0, τ_0, a_0, and, using the streamline model of the projectile, we find that the projectile flight can be described by the equation of motion of the projectile in standard atmosphere (air characteristics: ρ_{0N}, p_{0N}, τ_{0N}, and a_{0N}):

$$\frac{d\vec{v}}{dt} = \vec{g} - \overline{c} \cdot h_0(y) \cdot G_D(v)\frac{\vec{v}}{v}, \tag{2.6.1}$$

where

$$\overline{c} = \frac{p_0}{p_{0N}}\sqrt{\frac{\tau_{0N}}{\tau_0}} \cdot c, \tag{2.6.2}$$

with air density that changes with the ordinate y over the firing site according to the density function,

$$h_0(y) = (\frac{\tau_0 - 0.006328y}{\tau_0})^{4.4}, \tag{2.6.3}$$

where τ_0 is the virtual temperature at the firing site.

The origin of the coordinate system is at the departure point of projectile.

Thus, the systems of differential equations (2.1.3), (2.1.4), and (2.1.5) can be used to find the trajectory of flight by substituting in them (2.6.2) and (2.6.3).

Shooting at Sites with Low or High Temperatures

The vector differential equation (2.6.1) as well as the corresponding systems of the differential equations are still valid if we consider that the projectile is launched at the same firing site but with the atmospheric characteristics ρ_0, p_0, τ_0, and a_0 that are different from ρ_{0N}, p_{0N}, τ_{0N}, and a_{0N} (for example in a winter time, when the temperature at the sea level is too low, or in high temperatures of summer time).

We can use the PC programs ACA122.Bas and Ran122.Bas presented in section 2.4 to calculate the elements of the trajectory of flight in high-mountain shooting, or, in general, in any standard or non-standard atmosphere.

Example 6.1 Trajectory-Streamline Solution

The projectile caliber 122 mm is fired from a122mm Russian cannon, Mod 1960 with initial speed of 885m/s.
Use the PC program Aca122.bas to find the departure angle and all other elements of the trajectory at the impact point 12,000 meters from the cannon. The firing site and the target are located on the mountain at the altitude 2,000 meters over the sea level.
The temperature, the atmospheric pressure, and the pressure of water vapors are respectively 3 degrees Celsius, 589mmHg, and 2.9.
Consider that the temperature of the propellant is standard, 15 degrees Celsius.
At the sea level, the characteristics of the atmosphere are those of TSA.

Solution:

Input: x-coordinate of target, 12000; y-coordinate of target, 0; y-coordinate of cannon, 0; projectile departure speed, 885; x-coordinate of a point, 6000; range wind, 0; cross wind, 0; temperature of air, 3, temperature of propellant, 15; atmospheric pressure, 589; pressure of water vapors, 2.9.

Output: Departure angle, 7.2761 degree; time of flight, 19.79s; terminal speed, 434m/s; terminal angle, -11.678; trajectory vertex (6722m, 484m); ballistics coefficient, 0.2403; the fictive ballistics coefficient, 01929.

x-coordinate of a point: 6000m, corresponding y-coordinate, 477m; time, 8.12s; speed, 624; angle, 1.0878

Note:.

Actually, the departure angle given in the range tables of the respective cannon is 7.32 degrees.

Exercise 6.2

Solve the problem in example 6.1, if there is a range wind of -5m/s ("-:" sign shows that the wind is in the opposite direction of the x-axis), and there is a cross wind of 7m/s.

Answer:

Departure angle, 7.3218 degree; time of flight, 19.896s; terminal speed, 434m/s; terminal angle, -11.722; trajectory vertex (6735m, 489m); wind deflection, 43.6m; ballistics coefficient, 0.2403; the fictive ballistics coefficient, 01930.

x-coordinate of a point: 6000m, corresponding y-coordinate, 482m; time, 8.13s; speed, 627; angle, 1.089; wind deflection, 9.1m.

Example 6.3

The standard characteristics of the Russian caliber 122 mm-caliber projectile fired with the 122mm Russian cannon, Mod 1960 are:

- projectile speed, 885m/s, projectile mass, 27.3kg, temperature of propellant 15 degrees Celsius;

- TSA atmosphere: temperature, 15 degrees Celsius;, pressure, 750mmHg;, density of air, 1.205kg/m^3;; relative humidity 50% (pressure of water vapors, 6.35mmHg).

Use Ran122.Bas to find the range when a standard projectile is fired at an altitude of 1,500 meters over the sea level at an angle of 8.7 degrees. The temperature, the pressure, and the pressure of water vapors at the altitude of 1,500m are respectively 5.508 degrees Celsius, 625.9mmHg, and 3.4mmHg. The projectile speed is 885m/s.
Consider that the temperature of the propellant is 15 degrees Celsius.

Solution:

Input: y-coordinate of the firearm, 0; departure speed, 885; departure angle, 8.7, x-coordinate of a point, 0; range wind, 0; cross wind, 0; temperature of air, 5.508, temperature of propellant, 15; atmospheric pressure, 625.9; pressure of water vapors, 3.4

Output: Horizontal range, 13001m; time of flight, 22.91s; terminal speed, 392m/s; terminal angle, -14.9055; trajectory vertex (7404m, 651m); ballistics coefficient, 0.2422; the fictive ballistics coefficient, 0.2057.

Example 6.4

Use the PC program ACA122.BAS to find the departure angle of a Russian 122mm cannon Mod.1960 needed to hit the target located at the following ranges: 7,000m; 10,000m; 12,800m; 16,600m; 20,000m if the firing site is 1,500 meters over the sea level.
The temperature, the virtual temperature, and the pressure at the firing site are respectively: 5.508 degrees Celsius, 279.23 degrees Kelvin, and 625.90 mmHg. [16] The initial speed of projectile is 885m/s. The typical mass of the 122mm projectile is 27.3kg.

---

16.   "Mountain Shooting Tables of 122mm cannon Model 1960 ", Ministry of Defense of Albania, 1972.

Solution:

The fictive ballistics coefficient is

$$\bar{c} = \frac{p_0}{p_{0N}}\sqrt{\frac{\tau_{0N}}{\tau_0}} \cdot c = c \cdot \frac{625.9}{750} \cdot \sqrt{\frac{289.08}{279.23}} = 0.849125 \cdot c.$$

Note that the fictive BC is calculated automatically by the PC program.

Using the PC program ACA122.BAS for the range 16,600.

Input: x-coordinate of target, 16600; y-coordinate of target, 0; y-coordinate of gun, 0; departure speed, 885; temperature of air, 5.508; propellant temperature, 15; atmospheric pressure, 625.9; pressure of water vapors, 3.4; projectile standard mass, 2; change in mass, 0.

Output: Departure angle, 14.0907 degree; time of flight to the target, 34.03 seconds; terminal speed, 336m/s; terminal angle, -26.31966; coordinates of trajectory vertex, (9706m, 1461m); fictive ballistics coefficient, BC = 0.2128.

In a similar way, we find the elements of the trajectory for all other ranges. The results are displayed in table1.
In table 2 are displayed the elements of trajectory as they appear in the respective handbook of range tables of the given 122mm cannon. [11].

Table 1: Computed Trajectory Elements of a 122mm Projectile

Range x_T	Departure Angle	Time	Terminal Speed	Terminal Angle	Coordinates of Vertex	
7,000	3.4049	9.9	569	-4.57	3,789	120
1,0000	5.6232	15.7	472	-8.55	5,529	302
12,800	8.4604	22.4	397	-14.39	7,274	621
16,600	14.0907	34.0	336	-26.32	9,706	1,461
20,000	20.6448	47.2	326	-37.88	11,772	2,722

Table 2: Partial Data of Range Table, Projectile 122mm,
          Altitude 1500m

Range x_T	Departure Angle	Time	Terminal Speed	Terminal Angle	Coordinates of vertex
7,000	3.3833	9.8	575	-4.6	120
1,0000	5.6333	16.0	468	-8.5	301
12,800	8.4666	22.0	388	-15.0	616
16,600	14.1167	34.0	326	-27.0	1,480
20,000	20.8333	47.0	322	-39.0	2,800

Note:

The range table of the 122mm cannon is valid for the TSA atmosphere, i.e., at the sea level altitude or close to sea level (firing site), the temperature, the virtual temperature, and the pressure are respectively, 15 degrees Celsius, 289.08 degrees Kelvin, and 750 mmHg.

At the firing site 1,500 meters above the sea level, the temperature and pressure, given by the range tables, are respectively, 6 degrees Celsius and 626mmHg.

Example 6.5 Range Corrections

Range tables constructed for any projectile and the respective firearm include the so-called "range corrections." [17].

The shooting corrections are required to be known when the initial data of shooting are prepared using the standard range tables, but they are not necessary when we use the PC programs or other special programmable calculators that solve the differential equations that describe the flight of projectile to the target.

Firing range corrections are of two types (see chapter 4):

• Corrections in horizontal range caused by the deviation of the ballistics characteristics of the projectile and the characteristics of atmosphere from the standard characteristics.

[17.] Klimi, G., "Exterior Ballistics with Applications", Chapter 6, Xlibris, 2008.

- Correction in perpendicular direction of the firing plane caused by the cross wind, the projectile rotation, the rotation of the Earth (for relatively long -range shootings).

In this example, we show how we can estimate the *"range corrections"*.

The standard characteristics of the Russian caliber 122mm-caliber projectile fired with the 122mm Russian cannon, Mod 1960 are:

- projectile speed, 885m/s, projectile mass, 27.3kg, temperature of propellant 15 degrees Celsius.
- TSA atmosphere: temperature, 15 degrees Celsius; pressure, 750mmHg,; density of air, $1.205kg/m^3$; relative humidity 50% (pressure of water vapors, 6.35mmHg.)

For the horizontal range 12,000 meters in standard conditions, departure angle 8.6 degrees, *use PC programs Ran122.Bas or ACA122.BAS* to find the horizontal range corrections:

Correction in the direction of flight (deflection of the projectile in perpendicular direction, z-axis) that corresponds to a cross wind of 10m/s.

Range correction that corresponds to a change in range wind of 10m/s.

Range correction that corresponds to a change of 10mmHg in atmospheric pressure.

Range correction that corresponds to a change of 10 degrees Celsius in air temperature.

Range correction that corresponds to a change of 10 degrees in propellant temperature.

Range correction that corresponds to a change of 1% in departure speed.

Range correction that corresponds to a change of 0.67% in the projectile mass.

Range correction that corresponds to an increase of 0.06 degree in the angle of flight.

Solution:

Using Ran122.Bas:

To find the correction for each change, we input all standard data except the value that changes (as for example a change in pressure of 10mm, is entered in the input data as 760mm if the pressure is increased, and as 740 if the pressure is decreased).

Illustration: Correction that corresponds to a decrease/increase of 10 degrees Celsius in temperature (the actual temperature is respectively +5 degrees Celsius and 25 degrees Celsius).

Input: y-coordinate of firearm, 0; departure speed, 885m/s; departure angle, 8.6; temperature of air, 5; temperature of propellant, 15; pressure, 750; pressure of water vapors, 6.35; mass of projectile, 27.3; change in projectile mass, 0; range wind, 0; cross wind; 0; integration step, 0.5

Output: Range 11910 meters.

The range correction that corresponds to a decrease of 10 degrees Celsius in temperature is

$$\Delta x = 11910 - 12000 = -90m.$$

To find the correction in departure angle, we can use the ACA122.Bas. Inputting again the above data (temperature 5 degrees, range 12,000), we find that the departure angle is 8.726318. Thus the correction in departure angle is

$$\Delta \alpha = 8.7263 - 8.600 = 0.1263°.$$

In the same way, using Ran122.Bas, we find the following corrections:

Correction in range as result of a cross wind of 10m/s (deflection of the projectile in perpendicular direction of flight) is

$$\Delta x = 84m.$$

Correction in range as result of a range -wind of $+10$m/s (in direction of flight, along x-axis) is

$$\Delta x = 12125 - 12000 = 125m$$

Correction in range that corresponds to an increase in atmospheric pressure of 10mmHg is

$$\Delta x = 11927 - 12000 = -73m.$$

Correction in range that corresponds to an increase of 10 degrees in the propellant temperature is

$$\Delta x = 12157 - 12000 = 157m.$$

Correction in range that corresponds to a change 1% in the departure speed (-8.85m/s). Inputting departure speed 876.15 we find

$$\Delta x = 11845 - 12000 = -155m.$$

Correction in range that corresponds to a change of 0.67% of the projectile mass (increase in mass 0.183kg). Input mass, 27.3,; change in mass, 0.1836:

$$\Delta x = 11996 - 12000 = -4m$$

Change in range (correction in range) that corresponds to a change in the angle of departure of 0.06 degree.

Input: Departure angle 8.66:

$$\Delta x = 12045 - 12000 = 45m.$$

Note:

We can use the above value to find the *change in departure angle* (correction in departure angle) related with a certain change in range.

Illustration: Correction in departure angle that corresponds to a change of 8.85m/s in departure speed is

$$\Delta \alpha = 0.06 \cdot (-155) / (-45) = 0.20667° .$$

Thus, the departure angle should be approximately

$$\alpha = 8.6 + 0.20667 = 8.80667° .$$

Exercise 6.6 Range Corrections in High Mountains

Use the data of example 6.3 to find the correction in range that corresponds to an increase in temperature of air of 10 degrees if the shooting takes place at the altitude 1,500 meters over the sea level.

The normal values of the atmosphere at the altitude 1,500 meters are: temperature + 6 degrees Celsius,; pressure 626mmHg;, pressure of water vapors, 3.4mmHg. The propellant temperature is 15 degrees.

The departure angle needed to hit the target at the horizontal range 12,000m is 7.55 degrees.

Answer: 78m.

2.7. Snell's Law of Refraction Applied to the Projectile Flight

Following the postulate of Newton [18] on the corpuscular nature of wave and vice versa (Sir Isaac Newton describes light as small bullets), we find some simple original mathematical formulae that allow us to obtain the elements of the projectile trajectory in non-

---

18. Prof. Walter H. G. Lewin, Snell's Law; http://videolectures.net/mit802s02_lewin_lec29/ (web access on June -August, 2009)

standard atmosphere (high-altitude shootings, summer, or winter shootings) simply using the range tables already constructed for each firearm in standard atmosphere (ICAO, ASM, TSA).

No differential equations or ballistics coefficients are required to obtain the solutions that describe the projectile trajectory in non-standard atmosphere.

The model of a projectile-wave gives remarkably exact results solving the two main problems of exterior ballistics:

- Find the departure angle for given range.
- Find the range for a given departure angle.

The model of the projectile trajectory-streamline allows us to formally employ Snell's law of refraction to find the departure angle of a projectile (needed to hit a given target) as well as the other elements of the projectile trajectory at the terminal point or at any other point on the trajectory of flight.

Indeed, let's consider the projectile flight in non-standard atmosphere (medium 2) with density, pressure, virtual temperature, and sound speed respectively, ρ_0, p_0, τ_0, and a_0.

For the same projectile flying in standard atmosphere (medium 1), the density, the pressure, the virtual temperature, and the speed of sound are respectively, ρ_{0N}, p_{0N}, τ_{0N}, and a_{0N}.

We assume that the projectile trajectory-streamline has dynamics similitude described by the equation

$$\frac{1}{a_{0N}\rho_{0N}}\frac{dv_1}{dt} = \frac{1}{a_0\rho_0}\frac{dv_2}{dt}. \qquad (2.7.2)$$

Note that the flow field has dynamic similitude when the ratio of forces of the same nature that act in both fluid particles is constant.

Since the projectile mass is the same for both streamline trajectories, the equation (2.7.2) relates the trajectory streamline in non standard atmosphere with the trajectory streamline in standard atmosphere.

From (2.7.2) we can write:

$$\frac{dv_2}{dt} = \frac{\rho_0}{\rho_{0N}} \frac{a_0}{a_{0N}} \cdot \frac{dv_1}{dt} \tag{2.7.3}$$

where v_1 is the speed of projectile in standard atmosphere while v_2 is the speed of the same projectile in non-standard atmosphere.

The acceleration of the projectile in medium 1 is

$$\frac{dv_1}{dt} = c \cdot h(y) \cdot G_D(v_1), \tag{2.7.4}$$

where:

$$G_D(v_1) = (3.927 \times 10^{-4} \rho_{0N}) v_1^2 C_D (\frac{v_1}{a_{0N}}), \tag{2.7.5}$$

and

$$h(y) = (\frac{\tau_{0N} - 0.006328y}{\tau_{0N}})^{4.4}. \tag{2.7.6}$$

We denote v the speed of projectile in medium 2, i.e., $v = v_2$. Substituting in (2.7.3), we have:

$$\frac{dv}{dt} = \frac{\rho_0 a_0}{\rho_{0N} a_{0N}} \cdot \frac{dv_1}{dt}. \tag{2.7.7}$$

Projectile as a Point-Mass and "Wave"

We assume that the two atmospheres at the departure point of the projectile are in contact. At the fictive "interface surface" between two atmospheres, the "departure" angle of the projectile trajectory-streamline changes according to Newton's particle-wave interpretation of Snell's law.

Following the Newton's postulate that "a point-mass object manifests wave properties,", we must assume that, in absence of wind:

- the component of the velocity of the point-mass projectile along the fictive interface of two mediums remains constant;
- the initial speed of the projectile in each medium is equal to v_0.

To be compatible with the Newton's postulate that the component of the velocity along the interface remains unchanged, we assume that at the fictive interface-surface that separates two mediums:

- the component of initial velocity of the projectile along the interface remains constant, i.e., the magnitude of $\vec{v}$ does not change at the interface (figure 4);
- the projectile speed in the direction perpendicular to the interface changes abruptly form v_1 to v_2;
- the change of projectile speed in the direction perpendicular to the interface of two mediums causes the change in velocity, and as result, there is a change in the projectile acceleration that causes the projectile trajectory to deviate according to Snell's law (2.7.9).

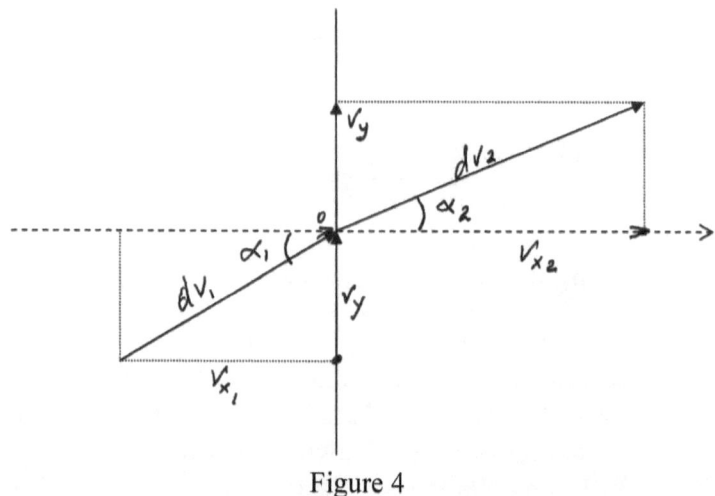

Figure 4

Thus, on both sides of the "interface," the change in velocity during the same interval of time dt is determined by the equation:

$$d\vec{v} = \vec{v} - \vec{v}_i.$$ (2.7.8)

From the geometry of figure 4, for the projectile trajectory - streamline we find:

$$\frac{\sin \alpha_2}{\sin \alpha_1} = \frac{dv_1}{dv_2},$$ (2.7.9)

where α_1 is the angle of projectile flying in standard atmosphere while α_2 is the corresponding similitude angle of the same projectile flying in non-standard atmosphere.

From (2.7.7) and (2.7.9), we obtain the following form of the Snell's law (2.7.9):

$$\frac{\sin \alpha_2}{\sin \alpha_1} = \frac{\rho_{0N} a_{0N}}{\rho_0 a_0}.$$ (2.7.10)

The parameter on the right side of Snell's law (2.7.10),

$$J = (\frac{\rho_{0N} a_{0N}}{\rho_0 a_0}),$$ (2.7.11)

is the ratio of the "characteristic acoustic impedances" of two mediums (atmospheres).

Using (2.7.10), for the departure angle of the projectile in the second atmosphere we obtain:

$$\alpha_{02} = ar \sin(\frac{\rho_{0N} a_{0N}}{\rho_0 a_0} \cdot \sin \alpha_{0N}).$$ (2.7.12)

Thus, the trajectory of the projectile flying in the non-standard atmosphere will have dynamics similitude with the given trajectory in the standard atmosphere if the departure angle α_{02} is given by (2.7.10), or (2.7.12).

What is the relationship between the other elements of the second trajectory (in non-standard atmosphere) and the corresponding elements of the "similitude" trajectory in the standard atmosphere?

We can apply the same model at any point on the second trajectory by associating to each point a fictive standard atmosphere with the properties that belong to the similitude point at the first trajectory streamline.

Note that since during the same trajectory the characteristics of atmosphere change continuously with the altitude over the sea level, according to Snell's law (2.7.10), the angle of refraction changes continuously from point to point.

The projectile angle (inclination) at any point on the second trajectory can be obtained using (2.7.10) and the associated projectile angle at the point of similitude on the first trajectory.

At the impact point, we can assume that the "terminal angle" of refraction is obtained using Snell's law (2.7.10) and the associated terminal angle at the similitude point on the first trajectory.

Consider the Siacci's solution of the differential equations of the projectile flight for high speeds of the projectile [19]:

$$x = -\frac{1}{Bg}[D(u) - D(v_0)]. \qquad (2.7.13)$$

Assuming that the initial speed and the terminal speed of the projectile on the second trajectory are respectively equal to the initial speed and the terminal speed of the projectile on the second trajectory, the analogy of the projectile wave-model with Siacci's formula (2.7.13), allows us to postulate that all the other elements of the second trajectory can be calculated using the "similitude" formulas:

$$x_2 = \frac{\rho_{0N} a_{0N}}{\rho_0 a_0} x_1, \qquad (2.7.14)$$

$$y_2 = (\frac{\rho_{0N} a_{0N}}{\rho_0 a_0})^2 y_1, \text{ *} \qquad (2.7.15)$$

---

19.   Klimi, G. "Exterior Ballistics with Applications", chapter 5, Xlibris, 2008.

$$t_2 = \frac{\rho_{0N} a_{0N}}{\rho_0 a_0} t_1 \qquad (2.7.16)$$

and

$$\alpha_{T2} = ar\sin(\frac{\rho_{0N} a_{0N}}{\rho_0 a_0} \sin\alpha_{T1}). \qquad (2.7.17)$$

The projectile speed in the second medium, at the point with coordinates

$$x_2 = \frac{\rho_{0N} a_{0N}}{\rho_0 a_0} x_1 \text{ and } y_2 = (\frac{\rho_{0N} a_{0N}}{\rho_0 a_0})^2 y_1 \qquad (2.7.18)$$

is equal to the speed of the projectile flying in the first medium at the similitude point with coordinates (x_1, y_1).

The y-coordinate in formula (2.7.15) is obtained as result of the superposition of two fields, the gravitational field and the air flow field, considering that:
Since the y-coordinate of the point mass projectile flying in free space is proportional to the square of the time of flight, we assume that the coefficient of proportionality of the ordinates of two similitude points is equal to the square of the quotient of characteristic acoustic impedances, i.e., proportional to:

$$J^2 = (\frac{\rho_{0N} a_{0N}}{\rho_0 a_0})^2. \qquad (2.7.19)$$

The formulas (2.7.14)-(2.7.18) can be used to find the range and all other elements of the second trajectory at the impact point when we know the corresponding elements at a similitude trajectory.

The coordinates of the vertex in the second trajectory are

$$x_{m2} = \frac{\rho_{0N} a_{0N}}{\rho_0 a_0} x_{m1} \text{ and } y_{m2} = (\frac{\rho_{0N} a_{0N}}{\rho_0 a_0})^2 y_{m1}. \qquad (2.7.20)$$

For the projectile flight, the Newton's interpretation of the Snell's law and the dynamics similitude of two trajectories show the following:

- The trajectory of a projectile in non-standard atmosphere is obtained from the trajectory of projectile in standard atmosphere using relations (2.7.14)-(2.7.20) and the Newton's interpretation of the Snell's law (2.7.10).
- The angle of departure when a projectile "passes through the fictive interface" between two media with different acoustics impedances (figure 4) changes according to Snell's law (2.7.10). That means that the departure angle of the projectile at a firing site, in non-standard atmosphere, is obtained using Snell's law and the corresponding angle of departure of projectile in standard atmosphere (or vice versa).
- The departure angle and the angle of projectile (inclination of the trajectory) at any point on the second trajectory change from the corresponding (similitude) point on the first trajectory.
- The projectile initial speed and the projectile terminal speed of the second trajectory are respectively equal to the initial speed and terminal speed of the same projectile flying in medium 1.
- The speed of projectile is the same for the points of two trajectories that have dynamics similitude.

Note that:

For the same departure angle, the ballistics coefficient that corresponds to trajectory 1 is equal to that of the trajectory 2 even though the range of the projectile changes according to the similitude laws.

As a matter of fact, it is the "fictive ballistics coefficient" that does change.

From the similitude relationships, it follows that the point on the trajectory 2 with coordinates

$$(x_2 = \frac{\rho_{0N} a_{0N}}{\rho_0 a_0} x_1, \quad y_2 = (\frac{\rho_{0N} a_{0N}}{\rho_0 a_0})^2 y_1), \qquad (2.7.21)$$

corresponds to the point with coordinates (x_1, y_1) on the trajectory 1.

At the same point, the angle of projectile flight is obtained from the first trajectory using the equation:

$$\alpha_2 = ar\sin(\frac{\rho_{0N} a_{0N}}{\rho_0 a_0} \cdot \sin\alpha_1). \qquad (2.7.22)$$

Practical Similitude Relations

For practical purposes, expressing the density and the speed of sound through pressure and virtual temperature, the relationships between the similitude points of two trajectories are obtained using the following relationships:

At the point on the second trajectory with coordinates

$$x_2 = \frac{p_{0N}}{p_0}\sqrt{\frac{\tau_0}{\tau_{0N}}} \cdot x_1 \quad \text{and} \quad y_2 = (\frac{p_{0N}}{p_0}\sqrt{\frac{\tau_0}{\tau_{0N}}})^2 y_1, \qquad (2.7.23)$$

the angle of flight α_2, the time of flight t_2, the projectile speed v_2, and the coordinates of the trajectory vertex are respectively given by the equations:

$$\alpha_2 = ar\sin(\frac{p_{0N}}{p_0}\sqrt{\frac{\tau_0}{\tau_{0N}}} \cdot \sin\alpha_1), \qquad (2.7.24)$$

$$t_2 = \frac{p_{0N}}{p_0}\sqrt{\frac{\tau_0}{\tau_{0N}}} \cdot t_1, \qquad (2.7.25)$$

$$v_2 = v_1 \qquad (2.7.26)$$

and

Projectile Trajectory

$$x_{m2} = \frac{p_{0N}}{p_0}\sqrt{\frac{\tau_0}{\tau_{0N}}} \cdot x_{m1}, \qquad y_{m2} = (\frac{p_{0N}}{p_0}\sqrt{\frac{\tau_0}{\tau_{0N}}})^2 \cdot y_{m1}. \qquad (2.7.27)$$

where α_1, t_1, and v_1 are respectively the angle, the time, and the projectile speed at the similitude point (x_1, y_1) located on the first trajectory.

Note that "similitude points" or "corresponding points" are two points (located on two different trajectories) whose coordinates satisfy the equation (2.7.23).

Since

$$J = \frac{p_{0N}}{p_0}\sqrt{\frac{\tau_0}{\tau_{0N}}}, \qquad (2.7.28)$$

we can write the above formulae as follows:

$$x_2 = J \cdot x_1, \qquad (2.7.29)$$

$$y_2 = J^2 y_1, \qquad (2.7.30)$$

$$t_2 = J \cdot t_1 \qquad (2.7.31)$$

and

$$\alpha_2 = ar \sin(J \cdot \sin \alpha_1). \qquad (2.7.32)$$

For the departure point and the terminal point, the last equation can be written respectively,

$$\alpha_{02} = ar \sin(J \cdot \sin \alpha_{01}) \qquad (2.7.33)$$

and

$$\alpha_{T2} = ar \sin(J \cdot \sin \alpha_{T1}). \qquad (2.7.34)$$

Application of Snell's Law and Similitude Laws in the Flight of Projectiles

The projectile-trajectory-streamline and Snell's law model allows us to obtain the elements of the second trajectory from the respective elements of the first trajectory and vice versa.

Thus, from the range tables that are prepared for a given projectile in standard atmosphere at the sea level (or close to it), we can obtain shooting data or the range tables of the same projectile flying in whatever atmosphere (high- or low-temperature atmospheres, or high-altitude atmosphere).

That is the well-known problem of firing practice related with the determination of the initial data of shooting in non-standard atmosphere using standard range tables.

Secondly, from the shooting tests performed in a non standard atmosphere, (winter, summer, or high-altitude atmosphere), we can obtain the range tables in standard atmosphere.

That is the well-known ballistics problem of "converting firing data to sea level standard data.".

The new approach is especially very useful in exterior ballistics of small arms to set up the departure angle using the individual data obtained in practice.

The following examples demonstrate the use of Snell's law and similitude laws to find the range of a projectile based on the range tables of a given projectile.

The formulas of similitude can be used as well to find the departure angle needed to hit a target located at an inclined range.

Comments:

- As a matter of fact, the similitude equations (2.7.14)-(2.7.18) express the relationships that exist between the elements of two similitude trajectories (corresponding trajectories) that have different departure angles but respectively identical initial speeds and identical terminal speeds.

- Though we can not talk for a real situation where the projectile trajectory -streamline, at the interface between two different mediums in contact, is refracted, i.e., although we do not have a real situation like in the case where the sound waves or the light waves are refracted passing through the interface of two mediums, the Newton's interpretation of the Snell's law for "refraction" of the projectile trajectory at the fictive interface gives valid results.
- The "wave behavior" of the projectiles and Newton's interpretation of the Snell's law applied to the projectile in flight give remarkably good results for the solution of the problems of exterior ballistics.

The following illustrations demonstrate the validity of Snell's law and the similitude model.

The accuracy of the Snell's law applied to the trajectory of flight is verified in the following section as well as in chapter 4.

Projectile Trajectory-Streamline and Snell's Law Model Applied to the Projectile Trajectory

The trajectory of a projectile is uniquely determined by the initial speed v_0, the departure angle α_0 and the ballistics coefficient c as well as by the characteristics of the atmosphere at the firing site.

The equation (2.7.3) that can be written in the following form,

$$\frac{dv_2}{dt} = \frac{\rho_2}{\rho_1} \frac{a_2}{a_1} \cdot \frac{dv_1}{dt},$$ (2.7.35)

allows us to deduce and use the Snell's law (according to Newton) [20] for the projectile trajectory assuming that:

- The projectile is launched in a standard atmosphere with density ρ_1, pressure p_1, virtual temperature τ_1 and speed of sound a_1.

20. Prof. Walter H. G. Lewin, Snell's Law; http://videolectures.net/mit802s02_lewin_lec29/ (web access on June-August, 2009)

- Immediately after the departure, the projectile crosses the interface that divides the standard atmosphere (the first atmosphere) from the non-standard atmosphere (or second atmosphere) with density p_2, pressure p_2, virtual temperature τ_2 and speed of sound a_2.

- Since the equation (2.7.35) does not tell us anything about the particular motion of the projectile or what it describes, we can say that the projectile at the launching point can be considered as composed by a flow of point mass projectiles with density ρ_i ($i = 1,2$) launched at a velocity $\bar{v}_i$ (speed v_0, departure angel α_i).

- At the interface the projectile stream rotates (is refracted) changing the direction of flow from α_1 to α_2.

Let's find the relationship between the departure angles α_1 and α_2.

The equation of conservation of mass for the fluid flow (continuity equation) is

$$\rho_1 dv_1 \cos\alpha_1 = \rho_2 dv_2 \cos\alpha_2. \qquad (2.7.36)$$

where ($dv_1 \sin\alpha_1$) and ($dv_2 \sin\alpha_2$) are respectively the component of the velocity along the x-axis.

The equations of the momentum conservation along the x-axis and y-axis are respectively

$$p_1 + \rho_1(dv_1 \cos\alpha_1)^2 = p_2 + \rho_2(dv_2 \cos\alpha_2)^2, \qquad (2.7.37)$$

and

$$\rho_1(dv_1 \cos\alpha_1)(dv_1 \sin\alpha_1) = \rho_2(dv_2 \cos\alpha_2)(dv_2 \sin\alpha_2). \qquad (2.7.38)$$

From (2.7.36) and (2.7.38) it follows that the component of the velocity change along y-axis remains constant, i.e.

$$dv_1 \sin\alpha_1 = dv_2 \sin\alpha_2. \qquad (2.7.39)$$

Hence we have

$$\frac{dv_2}{dv_1} = \frac{\sin \alpha_1}{\sin \alpha_2}. \tag{2.7.40}$$

From (2.7.35) and (2.7.40) we obtain the Snell's law (2.7.10), that can be written in the form:

$$\frac{\sin \alpha_2}{\sin \alpha_1} = \frac{\rho_1 a_1}{\rho_2 a_2}. \tag{2.7.41}$$

We would say that Snell's law (27.41), as it is interpreted by Newton, is valid for particles and it is in conformity with the laws of conservations of mass, momentum and energy though it is accepted that Snell's law is strictly a property of waves, and particularly of light.

Example 7.1 High-Altitude Shooting

Use the range tables of the Russian 122mm cannon Mod.1960 [21], [22] to find the departure angle needed to hit the target located at a range of 10,000 meters if the firing site is 1,500 meters over the sea level.

The initial speed of projectile is 885m/s.

The mountain range tables of the 122mm cannon are obtained for the TSA atmosphere. At the sea level, the temperature, humidity, and pressure are respectively, 15 degrees Celsius, 50%, and 750 mmHg.

Solution:

Employing

$$x_2 = \frac{p_{0N}}{p_0} \sqrt{\frac{\tau_0}{\tau_{0N}}} \cdot x_1$$

---

[21]. Range Tables of Cannon 122mm, Mod. 1960, Ministry of Defense of Albania, Tirana, 1967

[22]. Mountain Range Table of 122mm cannon Mod. 1960—Projectile OF-472, Ministry of Defense of Albania, Tirana 1972.

we find that the range of projectile in standard atmosphere is

$$x_{T1} = \frac{p_0}{p_{0N}}\sqrt{\frac{\tau_{0N}}{\tau_0}} \cdot x_{T2}.$$

Using the data of table 1, section 2.2, we find that the virtual temperature at the sea level and at the altitude 1,500 meters is respectively, $e_{0N} = 6.35$ and $e_0 = 3.536$.

The virtual temperature at the sea level and at the altitude 1,500 meters (over the sea level) is respectively:

$$\tau_{0N} = \frac{T_{0N}}{(1 - 0.3785 \cdot e_{0N} / p_{0N})} = \frac{(273.15 + 15)}{(1 - 0.3785 \cdot 6.35 / 750)} = 289.08$$

and

$$\tau_0 = \frac{T_0}{(1 - 0.3785 \cdot e_0 / p_0)} = \frac{(273.15 + 6)}{(1 - 0.3785 \cdot 3.536 / 626)} = 279.748.$$

Using the above data, we find that the corresponding similitude range of the projectile on the standard trajectory is at the horizontal distance

$$x_{T1} = \frac{p_0}{p_{0N}}\sqrt{\frac{\tau_{0N}}{\tau_0}} \cdot x_{T2} = \frac{626}{750}\sqrt{\frac{289.08}{279.748}} \cdot x_{T2} = 0.848468 \cdot (10,000) = 8484.68m \cdot$$

Departure Angle

Using the range table of the 122mm cannon (sea level range tables), by interpolation, we find that the departure angle needed to hit the target located at the horizontal range 8484.68 meters is $\alpha_{01} = 4.7872°$.

For J we have:

$$J = \frac{p_{0N}}{p_0}\sqrt{\frac{\tau_0}{\tau_{0N}}} = \frac{750}{626}\sqrt{\frac{279.748}{289.08}} = 1.178586 \, .$$

Substituting in Snell's law, we find that the departure angle (at the mountain site, 1,500 meters over the sea level) is

$$\alpha_{02} = ar\sin(J \cdot \sin\alpha_{01}) = \arcsin(1.178586 \cdot \sin(4.7872)) = 5.6447° \, .$$

The mountain range table of 122mm cannon referred above shows that:

> the departure angle that corresponds to the horizontal range 10,000 meters at the altitude 1,500 meters is $\alpha_2 = 5.63333°$

Other Elements of the Trajectory

The range table of a Russian 122mm projectile flying in standard atmosphere, for the range 8,484.68 meters, shows that that the time of flight, the terminal angle, the maximum altitude, and the terminal speed are respectively:

$$t_{T1} = 13.42s \, , \quad \alpha_{T1} = -7.2694° \, , \quad y_{m1} = 216.93m \, , \quad v_{T1} = 467.34m/s \, .$$

Using the similitude formulas, we find:

$$t_{T2} = J \cdot t_{T1} = 1.1786 \cdot (13.42) = 15.82s \, ,$$

$$\alpha_{T2} = ar\sin(1.1786 \cdot \sin(-7.2694)) = -8.5766° \, ,$$

$$y_{m2} = (1.1786)^2 \cdot (216.93) = 301.33m$$

and

$$v_{T2} = v_{T1} = 467.34m/s \, .$$

Note that at the altitude 1,500 meters, range 10,000 meters, the mountain range table of the Russian cannon 122mm shows that:

$t_{T2} = 16s$, $\alpha_{T2} = -8.5°$, $y_{m2} = 301m$, $v_{T2} = 468m/s$.

Employing the PC program ACA122.BAS for the range 10,000 meters at the altitude 1,500 meters we find:

$t_{T2} = 15.72s$, $\alpha_{T2} = 8.634°$, $y_{m2} = 304m$, $v_{T2} = 468.16m/s$

In table 1, there are displayed the results obtained above.

Table 1:—Elements of the trajectory: Range 10,000m

Method	α_{02}	Time	Angle	Altitude	Speed
Similitude-Snell	5.645	15.82	-8.577	301	467
Range Table	5.633	16	-8.8	301	468
PC Program	5.650	15.72	-8.63	304	468

Example 7.2 Estimation of the Departure Angle using Snell's Law

Use the table of the Russian 122mm cannon Mod.1960 (compiled for shooting at the sea level in TSA standard atmosphere) to find the departure angle needed to hit the target located at the following ranges: 7,000m,; 12,800m, 16,200m, 17000, 18000m, 20,600m, and 24,000 at the firing site at the altitude 1,500 meters.

The temperature, the virtual temperature, and the pressure at the firing site (1,500 meters over the sea level) are respectively,: 6 degrees Celsius (279.748 degree Kelvin) and 626 mmHg. The initial speed of the projectile is 885m/s.

The range table of the 122mm cannon is obtained for the TSA atmosphere: temperature, virtual temperature, and pressure at the sea level are respectively 15 degrees Celsius, 289.08 degrees Kelvin, and 750 mmHg.

Solution:

Using the range table of the 122mm projectile for the trajectory at the sea level, in standard atmosphere, and the same procedure as in example 7.1, hereafter are shown the elements of the trajectory for the given ranges:

(a)   Range $x_{T2} = 7000m$.

At the sea level, the similitude range is $x_{T1} = 5939.32m$.

The other elements of the trajectory at the similitude range and at the sea level are:

$\alpha_{01} = 2.9095°$, $t_{T1} = 8.08s$, $\alpha_{T1} = -3.7393°$, $y_{m1} = 86m$, $v_{m1} = 575.73m/s$.

Corresponding range at the altitude 1500 meters is $x_{T2}=7000m$. The corresponding elements of the trajectory at 1500 meters are:

$\alpha_{02} = 3.4297°$, $t_{T2} = 9.52s$, $\alpha_{T2} = -4.4083°$, $y_{m2} = 119.45m$, $v_{m2} = 575.73m/s$.

(b)   Range $x_{T2} = 12800m$

At the sea level, the similitude range is $x_{T1} = 10860.47m$.
The other elements of the trajectory at the similitude range and at the sea level are:

$\alpha_{01} = 7.1539°$, $t_{T1} = 19.30s$, $\alpha_{T1} = -12.30°$, $y_{m1} = 444.86m$, $v_{m1} = 387.19m/s$

Corresponding range at the altitude 1,500 meters is $x_{T2} = 12800m$. The corresponding elements of the trajectory at 1,500 meters are:

$\alpha_{02} = 8.4400°$, $t_{T2} = 22.75s$, $\alpha_{T2} = -14.5411°$, $y_{m2} = 617.94m$, $v_{m2} = 387.19m/s$

(c)   Range $x_{T2} = 16200m$

At the sea level, the similitude range is $x_{T1} = 13,745.28m$.

The other elements of the trajectory at the similitude range and at the sea level are:

$\alpha_{01} = 11.3709°$; $t_{T1} = 27.72s$, $\alpha_{T1} = -21.7264°$; $y_{m1} = 967.87m$, $v_{m1} = 324.57m/s$.

Corresponding range at the altitude 1,500 meters is $x_{T2} = 16,200m$. The corresponding elements of the trajectory at 1,500 meters are:

$\alpha_{02} = 13.4366°$, $t_{T2} = 32.68s$, $\alpha_{T2} = -25.8670°$, $y_{m2} = 1344.439m$, $v_{m2} = 324.57m/s$.

(d)   Range: $x_{T2} = 17000m$

At the sea level, the similitude range is $x_{T1} = 14424.06m$.

The other elements of the trajectory at the similitude range and at the sea level are:

$\alpha_{01} = 12.5941°$; $t_{T1} = 30.12s$, $\alpha_{T1} = -24.1203°$; $y_{m1} = 1147.22m$, $v_{m1} = 317.88m/s$.

Corresponding range at the altitude 1,500 meters is $x_{T2} = 17000m$. The corresponding elements of the trajectory at 1,500 meters are:

$\alpha_{02} = 14.8911°$, $t_{T2} = 35.50s$, $\alpha_{T2} = -28.7922°$, $y_{m2} = 1593.56m$, $v_{m2} = 317.88m/s$.

(e)   Range $x_{T2} = 18000m$.

At the sea level, the similitude range is $x_{T1} = 15272.53m$.

The other elements of the trajectory at the similitude range and at the sea level are:

$\alpha_{01} = 14.2284°$; $t_{T1} = 33.36s$, $\alpha_{T1} = -28°$; $y_{m1} = 1405.40m$, $v_{m1} = 314.54m/s$.

Corresponding range at the altitude 1,500 meters is $x_{T2} = 18000m$. The corresponding elements of the trajectory at 1,500 meters are:

$\alpha_{02} = 16.8389°$, $t_{T2} = 39.36$, $\alpha_{T2} = -33.5946°$, $y_{m2} = 1952.17m$, $v_{m2} = 314.64m/s$.

(e)   Range $x_{T2} = 20600m$

At the sea level, the similitude range is $x_{T1} = 17478.57m$.

The other elements of the trajectory at the similitude range and at the sea level are:

$\alpha_{01} = 18.9732°$, $t_{T1} = 41.39s$, $\alpha_{T1} = -36.3928°$; $y_{m1} = 2245.36m$, $v_{m1} = 314m/s$.

Corresponding range at the altitude 1,500 meters is $x_{T2} = 20600m$. The corresponding elements of the trajectory at 1,500 meters are:

$\alpha_{02} = 22.5314°$, $t_{T2} = 48.79s$, $\alpha_{T2} = -44.3690°$, $y_{m2} = 3118.95m$, $v_{m2} = 314m/s$.

(e)   Range $x_{T2} = 24000m$

At the sea level, the similitude range is $x_{T1} = 20363.38m$.

The other elements of the trajectory at the similitude range and at the sea level are:

$\alpha_{01} = 26.9007°$, $t_{T1} = 54.82s$, $\alpha_{T1} = -47°$, $y_{m1} = 3884.37m$, $v_{m1} = 322m/s$.

Corresponding range at the altitude 1,500 meters is $x_{T2} = 24000m$. The corresponding elements of the trajectory at 1,500 meters are:

$\alpha_{02} = 32.2251°$, $t_{T2} = 64.61s$, $\alpha_{T2} = -59.5378°$, $y_{m2} = 35395.64m$, $v_{m2} = 322m/s$.

In the following table are given for comparison the values of the departure angle taken from the range table of 122mm caliber projectile.

Table 1: Range, Departure Angle at 1,500 m Altitude.

Range	7000	10000	12800	16200	17000	18000	20600
α_{02} (Similitude)	3.430	5.645	8.440	13.437	14.891	16.839	22.53
α_{02} (Table)	3.383	5.633	8.467	13.433	14.833	16.667	22.2
Difference	0.047	0.012	-0.027	0.004	0.058	0.172	0.33
Relative Change	1.4%	0.21%	-0.3%	0.03%	0.39%	1.03%	1.4%

In table 2, we compare the results obtained using the Snell's law with the results obtained using the PC program ACA122.Bas, the range table, and the PC program Ballistica2.2. [23]

Table 2: Range, Departure Angle at 1500m Altitude

Range	7000	0000	12800	16200	17000	18000	20600
α_{02} (Similitude)	3.430	.645	8.440	13.437	14.891	16.839	22.53
α_{02} (ACA122)	3.404	.621	8.455	13.399	14.776	16.587	21.97
Ballistica2.2	3.390	.645	8.509	13.435	14.878	17.019	22.51
Range Table	3.383	.633	8.467	13.433	14.833	16.667	22.2

The above values of the departure angle are obtained using the PC program Ballistica2.2 and the G-function of resistance of the year 1943, employing the ballistic coefficient given in table 3: [24]

Table 3: The BC of 122mm Projectile, Departure Speed 885m/s.

Angle, α_0	$\leq 5°$	15°	25°	35°	45°
$c=f(\alpha_0)$, m²/kg	0.496	0.514	0.516	0.519	0.521
$C = 1.4222 / c$, lb/in²	2.867	2.767	2.756	2.740	2.730

23. Jan Krčmář, Ballistica2.2, http://www.balistika.cz/eng/exterior.html (accessed on 6 November, 2009)
24. "Firing Tables of 122mm cannon Model 1960 ", p. 117, Ministry of Defense of Albania, 1967

Since the BC is a function of the departure angle, we use the interpolation procedure to find the BC that corresponds to the intermediate values of the departure angle that are not presented in table 3.

Example 7.3 Range Calculation

The firing site of a caliber 122mm-caliber projectile is at an altitude 1,500 meters over the sea level.

The temperature, the virtual temperature, and the pressure at the firing site (1,500 meters over the sea level) are respectively: 5.508 degrees Celsius; 279.23 degrees Kelvin, and 625.90 mmHg. [25]. The initial speed of the projectile is 885m/s.

The range table of the 122mm cannon at the sea level shows that the horizontal range of 17,000 meters will be reached if the departure angle of the projectile is 17.8667 degrees.

What will be the horizontal range if the same departure angle will be set up at the firing site 1,500 meters over the sea level?

Solution:

Consider the trajectory of the projectile fired at an angle of 17.8667 at the altitude 1,500 meters over the sea level.

Using Snell's law, for the departure angle that corresponds to the similitude trajectory at the sea level, we have

$$\alpha_{01} = ar\sin(\frac{p_0}{p_{0N}}\sqrt{\frac{\tau_{0N}}{\tau_0}} \cdot \sin\alpha_{02}) = ar\sin(0.849125 \cdot \sin(17.8667)) = 15.1006°$$

Using the sea level range table of 122mm cannon, we find that the horizontal range that corresponds to the calculated above angle is 15667m.

The horizontal range at the altitude 1,500 meters is

$$x_{T2} = \frac{p_{0N}}{p_0}\sqrt{\frac{\tau_0}{\tau_{0N}}} \cdot x_{T1} = 1.17768(15667) = 18451m.$$

---

25.   "Mountain Range Tables of 122mm Projectile, Model 1960", Ministry of Defense of Albania, 1972

Example 7.4

Use the projectile trajectory-streamline and Snell's law model and the data in example 7.3 to find the horizontal range of a 122mm projectile fired with an angle of 12.1833 degree at an altitude 1,500 meters. Departure speed of projectile is 885m/s.

Note:
To find the similitude elements of the projectile trajectory at sea level shooting, use the PC program Ran122.Bas instead of the range table of the 122mm cannon.

Solution:

$$\alpha_{01} = ar\sin(\frac{p_0}{p_{ON}}\sqrt{\frac{\tau_{ON}}{\tau_0}} \cdot \sin\alpha_{02}) = ar\sin(0.849125 \cdot \sin(12.1833)) = 10.3231° \cdot$$

Using Ran122, we find that the corresponding horizontal range at the sea level is

$$x_1 = 13141m.$$

The horizontal range at the altitude 1,500 meters is

$$x_2 = J \cdot x_1 = 1.17768(13141) = 15476m.$$

Example 7.5

The test firing site is 3,000 meters over the sea level. The temperature and the pressure are respectively -3 degrees Celsius, and 520mm Hg.

Firing tests have shown that the departure angle needed to hit a target located at a horizontal range of 15,000 meters from a 122mm cannon Mod. 1960 is 21.36 degree.

Find the corresponding similitude range of the target at the sea level and the departure angle needed to hit it. In other words, we have to convert the firing data to the sea level considering the experimental data.

The departure speed of projectile is 885m/s.

Solution:

The departure angle of the sea-level similitude trajectory that corresponds to the departure angle of the same projectile but at the altitude 3,000m is:

$$\alpha_1 = ar\sin(\frac{p_0}{p_{0N}}\sqrt{\frac{\tau_{0N}}{\tau_0}}\cdot\sin\alpha_2) = \frac{520}{750}\sqrt{\frac{289.08}{270.15}}\sin(9.9) = 7.08317°\,.$$

The horizontal range to the target at the sea level is

$$x_1 = (\frac{p_0}{p_{0N}}\sqrt{\frac{\tau_{0N}}{\tau_0}}\cdot x_2) = \frac{520}{750}\sqrt{\frac{289.08}{270.15}}(15000) = 10{,}758.21m\,.$$

2.8. Snell's Law and Accurateness of Trajectory Calculations

For shooting at an altitude of 1,500 meters over the sea level, the values of the departure angle obtained in examples 7.1 and 7.2 using the trajectory similitude law and the Snell's law are within the errors of the range tables.

Indeed, the accuracy of range tables constructed by the middle of the 20th century is characterized by an average probable error E of 0.5% of the range of fire, [26], i.e.,

$$E_x = 0.005\cdot x_T\,, \tag{2.8.1}$$

and an average probable error in the departure angle of 2 minutes (0.0333 degree) that corresponds to a probable error of 0.0011 of the range, i.e.,

$$E_{2x} = 0.0011\cdot x_T\,. \tag{2.8.2}$$

---

26. Shapiro, J, M, "Exterior Ballistics", page 289 - 291, Oborongiz, 50'

Note that the probable error E is expressed through the standard deviation σ by the relation

$$E = 0.6745 \cdot \sigma.$$

Using table 1 of example 7.2, for the distance of 18,000 meters, we find that the probable error in the departure angle is

$$\Delta x_T / x_T = (0.0011 \cdot \Delta \alpha_0 / 0.0333) = (0.11) \cdot 0.186 / 0.0333)\% = 0.61\%,$$

while for the distance of 17,000 meters, the probable error is

$$\Delta x_T / x_T = (0.0011 \cdot \Delta \alpha_0 / 0.0333) = (0.11 \cdot 0.07 / 0.0333)\% = 0.23\%$$

Thus, till around 17,000-18000 meters, the probable error is within the probable error of the range tables. The probable error becomes bigger for ranges over 17,000-18,000 meters.

Maybe for ranges greater than 17,000 the probable errors of the range tables are larger than the average error $E_x = 0.005 \cdot x_T$ in accordance with the Shapiro's statement:

> "It is obvious that in some particular cases the real errors can be several times greater than the average error shown above; after the range tables are published the error becomes a systematic error.[27]

Another reason for the discrepancies of the results of table 1 (example 7.1) for large ranges can be derived from the fact that:

The ballistics coefficient with respect to the Russian law of resistance of the year 1943 that are used to estimate the elements of the trajectory are obtained (table 3) experimentally only for angles 5, 15, 25, 35, 45 degrees. [28]

---

27. Shapiro, J, M, "Exterior Ballistics", page 291, Oborongiz, 50'
28. "Range Tables of 122mm cannon Model 1960 ", p. 117, Ministry of Defense of Albania, 1967

Table 3: The BC of 122mm Projectile, Departure Speed 885m/s.

Angle, α_0	$\leq 5°$	15°	25°	35°	45°
$c = f(\alpha_0)$, m²/kg	0.496	0.514	0.516	0.519	0.521
$C = 1.4222 / c$, lb/in²	2.867	2.767	2.756	2.740	2.730

The results obtained in the following examples 8.1, and 8.2, using the trajectory similitude law and Snell's law, demonstrate once again the validity and the accurateness of the simple method of the trajectory similitude and Snell's law.

Notes:

In the practice of artillery, the accuracy of shooting is based on the standard range tables and is related with the number of rounds needed to be fired to hit the target located at a given range. The process requires the correction of firing range observing the impact point of a preceding round.

The better the predicted range, the smaller will be the number of rounds needed to correct firing and hit the target.

Note that in our calculations, we have neglected the jump (-2 minutes) of the 122mm cannon (projectile departure speed 885m/s) and have considered that the departure angle is equal to the elevation angle.

Example 8.1

In table 1, it is displayed the range table of the bullet 0.338 Lapua GB528 Scenar

19.44g (300 grain) fired with a muzzle speed of 830m/s, at the sea level in the ICAO atmosphere ((pressure 760 mm Hg, temperature 15 degrees Celsius, density of air 1.225kg/m³, speed of sound 340.30m/s, relative humidity 0%).

In table 2, it is given the range table of the same bullet fired at the altitude of 1,500 meters over the sea level (pressure 634.17mm Hg, temperature 5.25 degrees Celsius, density of air 1.058kg/m³, speed of sound 334.49m/s, relative humidity 0%).

Table 1 and table 2 are obtained using the PC program Ballistica 2.2 (English version 2.2) of Jan Krčmář, PhD. [29]

The bullet has a mass of 19.44 grams, and a diameter of 8.6mm. The form coefficient with respect to the drag function is 1, and the ballistics coefficient 3.8037.

Table 1: Range Table: Bullet 0.338 Lapua GB528 Scenar 19.44g;
        Sea Level

Range [m]	100	200	300	400	500	600
Initial Angle	0.0422	0.0873	0.1357	0.1877	0.2436	0.3037
Impact Speed	789	750	712	675	639	605
Time	0.124	0.254	0.392	0.536	0.688	0.849

Table 1 continued

Range [m]	700	800	900	1000	1100	1200
Initial Angle	0.3687	0.4389	0.5151	0.5977	0.6878	0.786
Impact Speed	572	539	508	478	449	422
Time	1.019	1.200	1.391	1.593	1.809	2.038

Table 1 continued

Range [m]	1300	1400	1500	2000	2500	3000
Initial Angle	0.8935	1.0112	1.1403	1.981	3.133	4.6068
Impact Speed	396	372	349	296	261	227
Time	2.282	2.543	2.822	4.402	6.210	8.290

Table 2: Range Table: 0.338 Lapua GB528 Scenar 19.44g; Altitude
        1,500m

Range [m]	200	300	400	600	800	1000
Initial Angle	0.087	0.1337	0.1839	0.2941	0.4193	0.5625
Impact Speed	761	728	696	634	576	521
Time	0.252	0.387	0.528	0.828	1.160	1.524

[29.] Jan Krčmář , Ballistica2.2 http://www.balistika.cz/eng/exterior.html (PC program downloaded on November 6th, 2009 and used with author's permission)

Table 2 continued

Range [m]	1200	1500	1800	2000	2500	3000
Initial Angle	0.7273	1.025	1.4015	1.7076	2.6745	3.1945
Impact Speed	470	399	339	315	282	253
Time	1.928	2.623	3.439	4.054	5.740	7.630

In table 3, there are displayed the elements of the trajectory of the same bullet when the firing site is 1,500 meters over the sea level, but the elements of the trajectory for each horizontal range are obtained using the similitude law and the Snell's law, i.e., using the following formulas:

$$x_2 = J \cdot x_1, \tag{1}$$

$$\alpha_{02} = ar\sin(J \cdot \sin\alpha_{01}), \tag{2}$$

$$v_2 = v_1, \tag{3}$$

$$t_2 = J \cdot t_1, \tag{4}$$

where

$$J = \frac{p_{0N}}{p_0}\sqrt{\frac{\tau_0}{\tau_{0N}}} = \frac{760}{634.17}\sqrt{\frac{278.4}{288.15}} = 1.17797.$$

For example, the elements of the trajectory for the horizontal range of 1,500 meters are obtained as follows:

Using (1), we find that the similitude horizontal range at the sea level is

$$x_1 = J^{-1} \cdot x_2 = (1.17797)^{-1} \cdot 1500 = 1273.40.$$

Using Jan's exterior ballistics PC program, we find that the elements of the trajectory at the sea level that correspond to the horizontal range 1273.40 are:

departure angle is 0.8639 degree, terminal speed is 403m/s, time of flight is 2.216.

Employing (2) and (4), we find that the departure angle and the time of flight that correspond to the range of 1,500 meters at the altitude of 1,500 meters are respectively:

$$\alpha_{02} = ar\sin(J \cdot \sin\alpha_{01}) = ar\sin(1.17797 \cdot \sin(0.78639)) = 1.0177°.$$

and

$$t_2 = J \cdot t_1 = 1.17797 \cdot (2.059) = 2.4254s.$$

The terminal speed is $v_{2T} = v_{1T} = 438m/s$.

In a similar way are obtained the elements of the trajectory displayed in table 3.

Table 3: Range Table Obtained Using Snell's Law, Altitude 1,500

Range [m]	200	300	400	600	800	1000
Initial Angle	0.0859	0.1332	0.1833	0.2932	0.4179	0.5594
Impact Speed	762	729	697	636	579	524
Time	0.252	0.385	0.525	0.827	1.157	1.520

Table 3 continued

Range [m]	1200	1500	1800	2000	2500	3000
Initial Angle	0.7229	0.8822	1.3891	1.6882	2.6331	3.8361
Impact Speed	473	403	344	320	287	258
Time	1.921	2.610	2.902	4.023	5.681	7.527

Comparing the results presented in table 3 with the results in table 2, we see that the estimation of the elements of the trajectory obtained using the similitude law and the Snell's law are accurate.

The differences in departure angle are noticeable over around 2,000 meters.

The results show that:

- At 2,000 meters, for example, the departure angle calculated using Snell's law is around 1 minute, and corresponds to a change in range about +10 meters, and to a change in vertical direction of 0.5 meters.

- For sniper shooting (1200-1500 meters), the results obtained using Snell's law give a very good accuracy. The error in the launching angle is 0.2-0.45 minutes and corresponds to a deviation in vertical direction of 1.6cm-17cm.
- Within the effective range of the small arms, 600-800 meters, the error in the launching angle is 0.03-0.08 minutes (0.4cm-1.6cm).

Comments:

The above errors in the estimation of the elements of the trajectory are obtained assuming that the radar measurements are very accurate, and that there are no errors in the solution method and the numerical integration of the differential equations of the projectile flight.

Since in general, the numerical integration introduces errors that are accumulated as the solution procedure evolves, I believe that the errors for large ranges reflect the accuracy of the solution method used to obtaining tables 1 and tables 2.

It is obvious that the errors that table 1 might contain are reflected in table 3 using the Snell's law.

Example 8.2 Projectile Drop Estimated Using Similitude Laws

In table 1, it is displayed the drop of the bullet 0.338 Lapua GB528 Scenar 19.44g (300 grain) fired horizontally with a muzzle speed of 830m/s, at the sea level in the ICAO atmosphere (pressure 760 mm Hg, temperature 15 degrees Celsius, density of air 1.225kg/m^3, speed of sound 340.30m/s, relative humidity 0%).

In table 2, it is given the drop of the same bullet fired horizontally at the altitude 2,200 meters over the sea level (pressure 581.53mm Hg, temperature 0.7 degree Celsius, density of air 0.986kg/m^3, speed of sound 331.74m/s, relative humidity 0%).

Table 1 and table 2 are obtained using the PC program Exterior Ballistics (English version 2.2) of Jan Krčmář Ph.D.

The bullet has a mass of 19.44 grams, and a diameter of 8.6mm. The form coefficient with respect to the drag function is 1, and the ballistics coefficient in SI units is 3.8037 m²/kg, (0.3739 lb/in²).

Table 1: Drop of Bullet 0.338 Lapua GB528 Scenar 19.44g; Sea Level

Range [m]	300	600	1200	1500	2000	2500	3000
Drop	-0.711	-3.182	-16.472	-29.878	-69.267	-13.71	-24.22
Impact Speed	712	605	422	349	296	261	228
Time	0.391	0.848	2.038	2.822	4.406	6.220	8.311

Table 2: Drop of Bullet 0.338 Lapua GB528 Scenar 19.44g; Altitude 2,200m

Range [m]	300	600	1200	1500	2000	2500	3000
Drop	-0.696	-3.038	-14.752	-25.698	-55.888	-108.4	-190.34
Impact Speed	735	647	491	423	329	292	265
Time	0.385	0.820	1.886	2.544	3.893	5.528	7.351

In table 3 is displayed the drop, the speed, and the time of flight of the same bullet when the firing site is 2,200 meters over the sea level. The elements of the trajectory for each horizontal range are obtained using the following formulas:

$$x_2 = J \cdot x_1 ,$$
(1)

$$y_2 = J^2 y_1$$
(2)

$$v_2 = v_1 ,$$
(3)

$$t_2 = J \cdot t_1 ,$$
(4)

where

$$J = \frac{p_{0N}}{p_0} \sqrt{\frac{\tau_0}{\tau_{0N}}} = \frac{760}{581.53} \sqrt{\frac{273.85}{288.15}} = 1.2740 .$$
(5)

For example, the elements of the trajectory for the range of 1,500 meters are obtained as follows:

Using (1), we find that the similitude horizontal range at the sea level is

$$x_1 = J^{-1} \cdot x_2 = (1.2740)^{-1} \cdot 1500 = 1177.34 \,.$$

Using Jan's exterior ballistics PC program Ballistics2.2, we find that the elements of the trajectory at the sea level that correspond to the horizontal range of 1,177.34 are:

projectile drop is -15.693, terminal speed is 428m/s, time of flight is 1.985.

Employing (2), (3), and (4), we find that the projectile drop, the terminal speed, and the time of flight that correspond to the range of 1,500 meters at the altitude of 2,200 meters are respectively:

$$y_2 = J^2 y_1 = 1.2740^2 \cdot (-15.693) = -25.47m \,,$$

$$v_2 = v_1 = 428m/s$$

and

$$t_2 = J \cdot t_1 = 1.2740 \cdot (1.985) = 2.529s \,.$$

Comparing the above values with the respective values of the elements of the trajectory presented in table 2 (for range 1,500 meters, altitude 2,200 meters: drop is -25.878, terminal speed is 349m/s, time of flight is 2.517 seconds), we find that they are approximately equal.

In a similar way are obtained the elements of the similitude trajectory that are displayed in table 3.

Table 3: Drop of Bullet 0.338 Lapua GB528 Scenar 19.44g; Snell's Law

Range [m]	300	600	1200	1500	2000	2500	3000
Drop	-0.694	- 3.029	-14.658	-25.47	-55.12	-106.2	-185.1
Impact Speed	736	650	495	428	336	299	271
Time	0.301	0.818	1.877	2.529	3.857	5.450	7.220

2.9. Langeven's Approach

The standard range tables of any firearm, at the sea level or in the proximity to the sea level, are valid for projectiles fired in a defined standard atmosphere (ASM, ICAO, TSA, etc.). They are accompanied with corrections related with small variations in the atmospheric characteristics at or near the sea level, or with small variations in the characteristics of the projectile from the accepted normal standard values.

The range tables contain as well the range corrections due to the influence of wind and other changes in the projectile characteristics (departure speed, projectile mass, temperature of the propellant charge, etc.) from the standard ones.

The range corrections for small variations of the characteristics of air at the sea level, especially for the artillery projectiles, can be obtained using (2.3.14), (2.3.15), or (2.3.16). Range corrections can be determined using different approaches as well as the numerical methods of integration of the differential equations of projectile flight or the Siacci's correction method. [30]

The sea-level altitude range tables for the ASM atmosphere, ICAO atmosphere, or TSA atmosphere can be obtained solving numerically the differential equations of the projectile flight (2.1.3), (2.1.4), or (2.1.5).

Using the PC programs obtained based on the differential equations of the projectile flight, we can solve in real time any exterior ballistics problem, prepare the range tables for a given firearm, or prepare the firing data to hit a given target.

Langeven's Approach

An effective approach that can be used to find the elements of the projectile trajectory, and so to construct the range tables when the temperature and the pressure at the launching point of the projectile are different from the normal standard values, is presented hereafter. It

30.   Klimi, G, "Exterior Ballistics with Applications", chapter 6, Xlibris, 2008

is based on the Langeven's formula, which is result of the integration of the differential equations of projectile flight (2.3.15). [31]

The Langeven's formula expresses the relationship that exists between the elements of two trajectories of the same projectile fired in different atmospheric conditions at the respective firing sites. The launching point of each projectile is not necessary located at the sea level.

Langeven's approach is based on the assumption that the density of air changes with altitude according to (2.1.6).

Consider the system of differential equation (2.3.15) that describes the flight of a projectile fired at a location, over the sea level, where the air temperature, τ_0, the air pressure, p_0, and the air density, ρ_0 are different respectively from the standard values: temperature, τ_{0N}, pressure, p_{0N}, and density, ρ_{0N} :

$$\begin{cases} \dfrac{dv_x}{dx} = -c \cdot \dfrac{p_0}{p_{0N}} h(y) \cdot \dfrac{G_D(v\sqrt{\tau_{0N}/\tau_0})}{v} \\ \dfrac{dp}{dx} = -\dfrac{g}{v_x^2} \\ \dfrac{dt}{dx} = \dfrac{1}{v_x} \\ \dfrac{dy}{dx} = p \end{cases} \quad ,(p = \tan\alpha) \quad (2.9.1)$$

where:

$$G_D(v\cdot\sqrt{\tau_{0N}/\tau_0}) = \begin{cases} A\cdot(v\sqrt{\tau_{0N}/\tau_0})^2 & for \quad v\cdot\sqrt{\tau_{0N}/\tau_0} \le 256m/s \\ E\cdot(v\cdot\sqrt{\tau_{0N}/\tau_0} - F & for \quad v\cdot(\tau_{0N}/\tau_0) > 256m/s \end{cases} ,$$

$$(2.9.2)$$

and

---

31.   Shapiro, J, M, "Exterior Ballistics", page 175, Oborongiz, 50'

$$h(y) = \left(\frac{\tau_{ON} - 0.006328y}{\tau_{ON}}\right)^{4.4}, \qquad (2.9.3)$$

Transforming (2.9.1), Langeven [32] has shown that the horizontal range of the projectile fired in a non standard atmosphere can be obtained using the system of differential equations (2.1.4), i.e.,

$$\begin{cases} \dfrac{dv_x}{dx} = -c \cdot h(y) \cdot \dfrac{G_D(v)}{v} \\[2mm] \dfrac{dp}{dx} = -\dfrac{g}{v_x^2} \\[2mm] \dfrac{dt}{dx} = \dfrac{1}{v_x} \\[2mm] \dfrac{dy}{dx} = p \end{cases} \qquad , \qquad (p = \tan \alpha)$$

$$h(y) = \left(\frac{\tau_{ON} - 0.006328y}{\tau_{ON}}\right)^{4.4},$$

by formally substituting in it:

- the ballistic coefficient c with the fictive ballistics coefficient

$$c^* = \frac{p_0}{p_{ON}} \cdot c; \qquad (2.9.4)$$

- the initial speed of the projectile v_0 with the fictive initial speed of the projectile

$$v_0^* = v_0 \cdot \sqrt{\tau_{ON} / \tau_0}. \qquad (2.9.5)$$

Thus, we obtain the following system of differential equations:

---

32.   Shapiro, J. M., "Exterior Ballistics", page 175, page 288, Oborongiz, 50'.

$$\begin{cases} \dfrac{dv_x}{dx} = -c^* \cdot h(y) \cdot \dfrac{G_D(v)}{v} \\[2mm] \dfrac{dp}{dx} = -\dfrac{g}{v_x^2} \\[2mm] \dfrac{dt}{dx} = \dfrac{1}{v_x} \\[2mm] \dfrac{dy}{dx} = p \end{cases} \quad , \quad (p = \tan\alpha) \qquad (2.9.6)$$

where

$$c^* = \frac{p_0}{p_{0N}} \cdot c, \quad v_0^* = v_0 \cdot \sqrt{\tau_{0N}/\tau_0}. \qquad (2.9.7)$$

Solving the system of differential equations (2.9.6), we find the elements of the trajectory y, v, t, $p=\tan\alpha$ at any point with abscissa x, or at point of impact on the ground (x_T, $y_T = 0$).

It can be shown that to find the range of the projectile in a given non-standard atmosphere, we multiply by the value of the quotient τ_0/τ_{0N} the range x_T that is obtained solving the system (2.9.6) for the initial speed (2.9.5). Thus, the actual range of the projectile in the given non-standard atmosphere is

$$x_T^* = \frac{\tau_0}{\tau_{0N}} x_T. \qquad (2.9.8)$$

The terminal speed, the time of flight, and the coordinates of the trajectory vertex are respectively:

$$v_T^* = v_T \cdot \sqrt{\tau_0/\tau_{0N}}, \qquad t_T^* = t_T \cdot \sqrt{\tau_0/\tau_{0N}}, \qquad (2.9.9)$$

$$((x_m^* = x_m(\tau_0/\tau_{0N}), \quad y_m^* = y_m(\tau_0/\tau_{0N})).$$

where v_T, t_T, x_m, y_m are obtained solving (2.9.6).

The above formulae are valid for any point on the trajectory. Thus, at a point with abscissa $x^* = \tau_0 x / \tau_{0N}$ we have:

$$y^* = y \cdot (\tau_0 / \tau_{0N}), \quad v^* = v \cdot \sqrt{\tau_0 / \tau_{0N}}, \quad t^* = t \cdot \sqrt{\tau_0 / \tau_{0N}}.$$
$$(2.9.10)$$

The ballistics coefficient (2.9.4) can be written:

$$c^* = \frac{\rho_0 a_0^2}{\rho_{0N} a_{0N}^2} \cdot c.$$

Transforming (2.9.10), we have:

$$y^* = \frac{a_0^2}{a_{0N}^2} y, \quad t^* = \frac{a_0}{a_{0N}} t, \quad v^* = \frac{a_0}{a_{0N}} v, \quad (2.9.11)$$

Since the Langeven's approach is obtained using the system of differential equations (2.9.1) it is obvious that the solution of the system of differential equations (2.9.6) is equivalent to the solution of the system (2.9.1). The validity of the above statement is illustrated in example 9.1.

Nowadays, it is convenient to use the general form of the differential equations of the projectile flight (2.9.1) instead of (2.9.6).

Notes:

One of the results of Langeven's approach is the Langeven's formula [33]

$$x = \frac{\tau_0}{\tau_{0N}} F(c^*, v_0^*, \alpha_0), \quad (2.9.12)$$

that can be used if there are known the ballistics tables,[34], (see section 2.14).

---

33.   Shapiro, J. M., "Exterior Ballistics", page 175, Oborongiz, 50'.

The Langeven's approach and the conclusions that can be drawn from the Langeven's formula are fundamental for the solution of exterior ballistics problems.

The Langeven's formula shows the following:

- The ballistics coefficient c of a given projectile measured for the projectile flight in a given standard atmosphere is the same for any atmosphere, when at the firing site the atmospheric conditions are different from the characteristics of the standard atmosphere at the sea level.
- To estimate the elements of the trajectory of a projectile, we can use the general differential equations of the projectile flight (2.3.14), (2.3.15) or (2.3.16) considering that the atmospheric characteristics of air are measured at the firing - site, assuming that the center of the Cartesian rectangular system of coordinates is at the launching point of the projectile.

Note that the following exercises are based on the 122mm Russian projectile fired by the 122mm field cannon, Mod. 1960 since actually I have access only to such range tables.

Langeven's Approach and the Projectile-Streamline Approach

Comparing the Langeven's approach with the projectile-streamline approach, we notice that:

- The Langeven's approach considers not only a fictive ballistics coefficient (2.9.4) but also a fictive departure speed of the projectile (2.9.5);
- the elements of the trajectory obtained using the Langeven's formula and the ballistics tables need to be corrected using two multipliers in order to find the actual elements of the trajectory;

[34.] De Mestre, N, "The Mathematics of Projectiles in Sport", p. 60, Cambridge University Press, 1990.
Edward J. McShane, John L. Kelly, Franklin Reno, "Exterior Ballistics", p. 246, The University of Denver Press, 1953.

- the projectile-streamline approach is convenient since it uses only the fictive ballistics coefficient (2.4.11) to calculate directly the elements of the trajectory;.
- using the Langeven's model, we need to use the ballistics tables to calculate the trajectory of a given projectile;
- for both models, we can solve the equations of projectile flight with fictive ballistics coefficients;
- the projectile-streamline approach is simple and has less mathematical operations than the Langeven's method.

Example 9.1

Find the range of a Russian 122mm projectile fired with an angle of 6.20 degrees and speed of 885m/s when the pressure and the temperature at the firing point located at the sea level are respectively 760mm and 25 degrees Celsius, humidity 50%, the pressure of water vapors is 11.88mm (see table 1 section 2.2) . The virtual temperature is 299.92. Ballistics coefficient is $c=0.2389$. The typical mass of the 122mm projectile is 27.3kg.

(a) Use the PC program RangeC.Bas [35] and the Langeven's method

(b) Use the PC program Rameco.Bas. [36]

Solution:

(a) The fictive ballistics coefficient is

$$c* = \frac{p_0}{p_{ON}} \cdot c = \frac{760}{750}c = 1.01333 \cdot (0.2389) = 0.2421$$

The fictive initial speed of the projectile is

$$v_0* = v_0 \cdot \sqrt{\tau_{ON} / \tau_0} = (885)\sqrt{289.08 / 299.92} = (885) \cdot (0.98175) = 868.85 .$$

---

35. Klimi, G. "Exterior Ballistics of Small Arms", PC Program RangeC.Bas, page 187 - 191, Xlibris, 2009
36. Klimi, G. "Exterior Ballistics of Small Arms", PC Program Rameco.Bas, page 216 - 220, Xlibris, 2009

Using RangeC.Bas

Input: y-coordinate of firearm, 0; Initial speed, 868.85; Departure angle, 6.2; Ballistics coefficient, 0.2421; Integration step, 1.

Output: Range, 9693; Time of flight, 16.47; terminal speed, 415.6; Impact angle, -10.1298 degree; Trajectory vertex (5453, 335.3).

Using the Langeven's approach, we find the following real values:
Range is

$$x_T{}^* = \frac{\tau_0}{\tau_{0N}} x_T = \frac{299.92}{289.08} \cdot (9693) = 10{,}057m.$$

Time of flight is

$$t^* = \sqrt{\frac{\tau_0}{\tau_{0N}}} \cdot t = \sqrt{\frac{299.92}{289.08}} \cdot 16.49 = 16.78s.$$

Terminal speed is

$$v_T{}^* = v_T \cdot \sqrt{\frac{\tau_0}{\tau_{0N}}} = 415.6\sqrt{\frac{299.08}{289.08}} = 423.32m/s.$$

The coordinates of the trajectory vertex are

$$x_m{}^* = \frac{\tau_0}{\tau_{0N}} x_m = \frac{299.92}{289.08} \cdot (5453) = 56557.5m,$$

$$y_m{}^* = \frac{\tau_0}{\tau_{0N}} y_m = \frac{299.08}{289.08} \cdot (335.3) = 348m$$

(b) Using PC program Rameco.Bas (*Exterior Ballistics of Small Arms*, page 214, Xlibris, 2009, page 214)

Input: y-coordinate of firearm, 0; Initial speed, 885; departure angle, 6.20; temperature of air, 25; Temperature of propellant, 15; Atmospheric pressure, 760; Pressure of air vapors,11.88; Projectile mass, 27.3; Change in projectile mass, 0; Range wind, 0; Cross wind, 0; Ballistics coefficient, 0.2389; Integration step, 1.

Output: Range, 10058m; Time of flight, 16.77; Terminal speed, 423m/s; Terminal angle, -10.13 degree; Trajectory vertex (5658m, 348m).

Note:

From the range table of the Russian 122mm projectile fired with 122mm cannon Mod.1960, we find that the projectile range is 10,053m.

2.10. Langeven's Approach for Shooting in High Mountains

Using the differential equations of projectile flight (2.3.14), (2.3.15), and (2.3.16), we can solve any exterior ballistics problem when we know the initial conditions of the projectile, the properties of air at the sea level, as well as the ballistics wind.

In practice, it is convenient to find the firing data using the measurements of the temperature, pressure, and humidity of air at the firing site.

According to Shapiro [37], the Langeven's approach can be used to find the range of a projectile fired in a relatively high altitude y over the sea level if we consider the characteristics of atmospheric air at the firing site as the characteristics of the second atmosphere.

Indeed, considering the conclusions obtained in section (2.3), we can use the system of differential equations (2.3.1), i.e.

---

37. Shapiro, J, M, "Exterior Ballistics", page 288, Oborongiz, 50'

$$\begin{cases} \dfrac{dv_x}{dx} = -c \cdot \dfrac{p_0}{p_{0N}} h(y) \cdot \dfrac{G_D(v\sqrt{\tau_{0N}/\tau_0})}{v} \\[2mm] \dfrac{dp}{dx} = -\dfrac{g}{v_x^2} \\[2mm] \dfrac{dt}{dx} = \dfrac{1}{v_x} \\[2mm] \dfrac{dy}{dx} = p \end{cases} \qquad , \quad (p = \tan\alpha) \qquad (2.5.1)$$

where

$$G_D(v\cdot\sqrt{\tau_{0N}/\tau_0}) = \begin{cases} A\cdot(v\sqrt{\tau_{0N}/\tau_0})^2 & for \quad v\cdot\sqrt{\tau_{0N}/\tau_0} \le 256m/s \\[3mm] E\cdot(v\cdot\sqrt{\tau_{0N}/\tau_0} - F & for \quad v\cdot(\tau_{0N}/\tau_0) > 256m/s \end{cases} ,$$

$$(2.5.2)$$

and

$$h(y) = (\frac{\tau_{0N} - 0.006328y}{\tau_{0N}})^{4.4} . \qquad (2.5.3)$$

to find the range of a given projectile launched at a high mountain, assuming that the projectile departs from the origin of the coordinates that is located at the muzzle of the firearm. As a matter of fact, we consider the second atmosphere the atmosphere (and its characteristics) at the altitude y.

As we demonstrate in Example 10.1, the range of fire estimated using the Langeven's approach is acceptable.

Employing the differential equations of projectile flight, we can solve any exterior ballistics problem, if we know (measure or calculate in advance) all the parameters that determine the projectile flight:

- the parameters related to the atmospheric characteristics (pressure, temperature, etc.);
- the nominal parameters related to the characteristics of the firearm (BC, initial speed, etc.);

- the deviation of the projectile and firearm characteristics from the nominal data (departure speed, mass of projectile, temperature of propellant, etc.), as well as the corrections related with the influence of wind (ballistics wind), Earth rotation, projectile rotations, etc., in the flight of the projectile.

The corrections related with the rotation of the projectile and the Earth can be found in the manuals of firearms, and then they can be introduced in the PC programs compiled at the end of the chapter.

The Langeven's approach is very useful for small arms shooting when we can measure the characteristics of atmosphere at the firing site.

Example 10.1

Find the range of a 122mm projectile fired with an angle of 6.20 degrees, and speed of 885m/s when the pressure and temperature at the firing point located at 1,500m over the sea level are respectively 625.9mm and 6 degrees Celsius. The ballistics coefficient is 0.2389.

(a) Use the PC program RangeC.Bas and the Langeven's method.
(b) Use the PC program Rameco.Bas .(see "*Exterior Ballistics of Small Arms*").

Solution:

(a) The fictive ballistics coefficient is

$$c^* = \frac{p_0}{p_{ON}} \cdot c = \frac{625.9}{750} c = 0.83453(0.2389) = 0.19937$$

The fictive initial speed of the projectile is

$$v_0^* = v_0 \cdot \sqrt{\tau_{ON} / \tau_0} = \sqrt{289.08 / 279.23} = (885) \cdot (1.01748\cdot) = 900.5 .$$

Using Rangec.Bas:

Input: y-coordinate of firearm, 0; Initial speed, 899.11; Departure angle, 6.2; Ballistics coefficient, 0.19949; Integration step, 1.

Output: Range, 10,968; Time of flight, 17.35; terminal speed, 458.02; Impact angle, -9.7114 degree; Trajectory vertex (6110m, 371.1m).

Using the Langeven's approach, we find the following real values:

Range

$$x_T{}^* = \frac{\tau_0}{\tau_{ON}} x_T = \frac{280.08}{289.08} \cdot (10968) = 10,626.5m .$$

Time of flight

$$t^* = t\sqrt{\frac{\tau_0}{\tau_{ON}}} = 17.35\sqrt{\frac{280.08}{289.08}} = 17.08s .$$

Terminal speed

$$v_T{}^* = \sqrt{\frac{\tau_0}{\tau_{ON}}} v_T = \sqrt{\frac{280.08}{289.08}} \cdot 458 = 451m / s .$$

Trajectory vertex

$$x_m{}^* = \frac{\tau_0}{\tau_{ON}} x_m = \frac{280.08}{289.08} \cdot (6110) = 5920m ,$$

$$y_m{}^* = \frac{\tau_0}{\tau_{ON}} y_m = \frac{299.08}{289.08} \cdot (371.09) = 360m .$$

(b) Using PC program Rameco.Bas

Considering the projectile altitude at 0m, instead of 1500m
Input: y-coordinate of firearm, 0; Initial speed, 885; departure angle, 6.20; temperature of air, 6; Temperature of propellant at firing site, 15; Atmospheric pressure at firing site, 626; Pressure of air vapors at firing site, 6.35; Projectile mass, 27.3; Change in projectile mass, 0; Range wind, 0; Cross wind, 0; Ballistics coefficient, 0.239; Integration step, 1.

Output: Range, 10625m; Time of flight, 17.08; Terminal speed, 451m/s; Terminal angle, -9.713 degree; Trajectory vertex (5919m, 360m).

Using the range table of the Russian 122mm projectile 122mm range table, we find that the projectile range is 1,0640m.

The results obtained in a, and b are approximate.

2.11. Theoretical Estimation of the Form Coefficient

The trajectory of a given point-mass projectile in flight is a function of projectile initial speed, departure angle, projectile form coefficient (ballistics coefficient), and the characteristics of the atmosphere.

We consider that the projectile speed is the standard speed given in the manufacturer's manual of the firearm and projectile, while the departure angle can be set up using the firearm mechanisms of sighting and the characteristics of shooting.

We need to determine the ballistics coefficient (that is present in the systems of differential equations of projectile flight),

$$c = \frac{id^2}{m}1000, \qquad (2.11.1)$$

where d is the caliber of the bullet (caliber of the firearm), m is the bullet mass, and i is the form coefficient.

Mainly in the USA and in the Great Britain, there is in use the definition of the ballistics coefficient:

$$C = \frac{m}{i \cdot d_e^2}. \qquad (2.11.2)$$

where d_e is the "reference diameter" (the diameter of the pre-flight bullet), m is the bullet mass, and i is the form coefficient.

The mass of the projectile and its caliber or reference diameter are equal to the standard values given in the manufacturer's manual.

In this book, we use the diameter of the projectile in flight that is approximately equal to the caliber of the firearm.

To be consistent with one or the other definition of the ballistics coefficient (and so with the results obtained solving the differential equations of projectile flight), in some problems we need to convert the value of the ballistics coefficient c into C (and vice versa) using the formula:

$$C = \frac{1.4222}{c} \frac{d^2}{d_e^2},$$

(2.11.3)

instead of [38]

$$C = \frac{1.4222}{c}.$$

(2.11.4)

Thus, for the firearm with aof caliber of 7.62mm (0.30 inches), the projectile reference diameter is 7.82mm (0.308 inches). A bullet that has a Siacci's ballistics coefficient $c=4.2$, has a corresponding ballistics coefficient

$$C = \frac{1.4222}{c} \frac{d^2}{d_e^2} = \frac{1.4222}{4.208} \frac{(7.62)^2}{(7.82)^2} = 0.3209$$

and not

$$C = \frac{1.4222}{c} = \frac{1.4222}{4.208} = 0.3380.$$

The best way to estimate the ballistics coefficient is the use of the experimental results of shooting with firearms and projectiles, as we have shown in "*Exterior Ballistics with Applications*" and in "*Exterior Ballistics of Small Arms*".

---

38. Klimi, G. "Exterior Ballistics with Applications", p. 80, Xlibris, 2008

The form coefficient depends on the particular drag coefficient (G-function) we use to study the projectile trajectory. For a given G-function, it depends on the projectile speed.

Anyway, to simplify the solution of the differential equations of the projectile flight, the traditional exterior ballistics considers the form coefficient of a given projectile as a parameter that depends on the departure angle, or, equivalently, on the firing range.

For the bullets, following R. McCoy, the modern exterior ballistics considers a form coefficient that is an average of some form coefficients estimated using Siacci's functions, or the sparking photography methods, or some rules based on the type of projectile and the nose length. [39].

In *"Exterior Ballistics of Small Arms"*, it is recommended to use a ballistics coefficient (form coefficient) that is measured for ranges, of 500 meters, 800 meters, 1,500 meters, etc.

The coefficient of form of a bullet that corresponds to the Siacci function can be estimated employing only the geometrical characteristics of the bullet based on Colonel Tzernozubovim's formula: [40]

$$i = \frac{0.3}{\sqrt{0.5+10d}} \cdot (1.23 - 0.15 \cdot \frac{d}{d'}) + \frac{h}{1+4h^2} \cdot \frac{12850+K}{4200} - \frac{500}{1200+v_0}(1-(\frac{d'}{d})^2),$$

$$(2.11.5)$$

where

- d is the caliber of the firearm in decimeters;
- d' is the diameter of the bottom part (base) of the projectile;
- h is the height of the ogive (head length) in calibers;
- v_0 is the initial speed of the projectile;
- the value of the parameter K is determined by the formula:

$$K = \begin{cases} x_T & \text{for } x_T < 1800 \cdot (d / 0.0762)^2 \\ 1800(d / 0.0762)^2 & \text{for } x_T \geq 1800 \cdot (d / 0.0762)^2 \end{cases}, \quad (2.11.6)$$

[39]. R. L. McCoy, "Modern Exterior Ballistics", p.106-p.111
[40]. Shapiro, J. M. "Vneshnaja Ballistica", p.59, Moscow 50'.

- x_T is the projectile distance to the point of impact.

The form coefficient estimated using formula (2.11.5) depends on the projectile range that is a function of the departure angle.

The form coefficient increases with the range, but remains constant for projectile ranges greater than

$$x_T \geq 1800 \cdot (d / 0.0762)^2. \tag{2.11.7}$$

The Tzernozubovim's formula (2.6.8) is not valid for the artillery projectiles.

The theoretical determination of the BC of a bullet using (2.11.5) needs to be verified in practice.

Anyway, it is helpful to have an idea about the BC of a given bullet that can be used for field experiments, especially for relaltively long ranges.

Example 11.1 Ballistics Coefficient of 0.50 Ball M33 Bullet

Find the Siacci form coefficient for the 0.50 Ball M33 projectile with the following characteristics [41]: $d = 0.0127m$, $d' = 0.009750m$, $h=0.033152m$.

The Projectile speed is $v_0 = 900m / s$. Mass of the bullet is $m = 0.0421kg$. The reference diameter is $d_e = 0.01295m$.

Estimate the form coefficient for the following distances: 300, 500m, 800m, 1,000m, 1,200m, 1,500m, 2,000m, 2,500m, 3,000, 3,500m, 4,000m.

Solution:

Bullet range is 500 meters. The value of the parameter K is 5000, since

$$x_T = 500 < 1800(d / 0.0762)^2 = 1800(0.127 / 0.0762)^2 = 5000. \tag{1}$$

Substituting $d = 0.127dm$, $d' = 0.09750dm$, $h=0.033152m=2.61039$ *calibers* and $v_0 = 900m / s$ in (2.1.15) we find that the form coefficient is $i=0.4314$.

[41.]   0.50 Ball M33,. http://www.dtic.mil/dticasd/sbir/sbir031/n005a.pdf

The ballistics coefficient that corresponds to Siacci G-function is

$$c = \frac{(0.4314) \cdot (0.01275)^2}{0.0421} 1000 = 1.6527 \,.$$

In English units, the ballistics coefficient is

$$C = \frac{1.4222}{c} = \frac{1.4222}{1.6527} = 0.8605 \,.$$

If we consider the reference diameter $d_e = 0.01295m$ instead of the caliber, we find that the ballistics coefficient is

$$C = \frac{1.4222}{c} \frac{d^2}{d_e^2} = \frac{1.4222}{1.6527} \frac{(0.0127)^2}{(0.01295)^2} = 0.8276 \,.$$

In the same way, we find the Siacci form coefficients and the respective ballistic coefficients for ranges 800m, 1000m, ... 4,500m, 5,000m, 5,500m, ... The values of the form coefficients and ballistics coefficient are displayed in table 1.

From the condition 1, it follows that for the range

$$x_T \geq 5000m \,,$$

the parameter K is constant, $K = 5000$.

We conclude that for ranges greater than or equal to $x_T \geq 5000m$, the form coefficient remains constant and equal to the value 0.4702 that corresponds to the range $x_T = 5000m$.

Note that we can to use that property of the Siacci form coefficient to estimate the maximum range of bullets.

Table 1: Projectile 0.50 Ball M33, Initial Speed of Projectile $v_0 = 900m/s$

x_T	100	200	300	400	500	600	700	800
i	0.4226	0.4248	0.4270	0.4292	0.4314	0.4336	0.4358	0.4380
c	1.6190	1.6275	1.6359	1.6443	1.6527	1.6612	1.6696	1.6780

x_T	900	1000	1100	1200	1300	1400	1500	1600
i	0.4402	0.4424	0.4446	0.4468	0.4490	0.4512	0.4534	0.4556
c	1.6865	1.6949	1.7033	1.7117	1.7202	1.7286	1.7370	1.7455
x_T	2000	2500	3000	3500	4000	4500	5000	...
i	0.4644	0.4753	0.4863	0.4973	0.5083	0.5193	0.5303	0.5303
c	1.7792	1.8209	1.8631	1.9052	1.9474	1.9895	2.0316	2.0316

Example 11.2 Sierra 0.308, Caliber 7.62mm, 168 Grain, MatchKing

Find the Siacci form coefficient for the Sierra 0.308 caliber, 168 grain MatchKing bullet with the following characteristics [42]: Caliber, $d = 0.00762m$; diameter of the bottom of the bullet, $d'=0.006147m$; height of the ogive, $h=0.017526m$; initial speed of projectile, $v_0=807.72m/s$. (Mass of the bullet is $m=0.0109kg$, and the reference diameter is $d_e=0.00782m$.

Estimate the form coefficient for the following distances: 100m, 200m, ... 1200m, 1500m, 1700, 1800m, 2000m.

Solution:

In the same way as in example 1, we find that for ranges greater than or equal to $x_T = 1800m$, the parameter K has a constant value, $K = 1800$.

Table 2: Sierra 0.308 caliber, 168 grain, MatchKing, $v_0=807.72m/s$

x_T	100	200	300	400	500	600	700	800
i	0.5118	0.5143	0.5168	0.5193	0.5217	0.5242	0.5267	0.5291
c	2.7300	2.7432	2.7565	2.7699	2.7827	2.7960	2.8093	2.8221
x_T	900	1000	1100	1200	1500	1700	1800	2000
i	0.5316	0.5341	0.5366	0.5390	0.5464	0.5514	0.5539	0.5539
c	2.8355	2.8488	2.8621	2.8750	2.9144	2.9411	2.9544	2.9544

---

[42]. Litz, B., Applied Ballistics for Long Range Shooting, Applied Ballistics, LLC, 2009

McDonald, W., Inclined Fire, June, 2003, http://www.exteriorballistics.com/ebexplained/5th/50.cfm

In table 3 is given the ballistics coefficient of the same bullet, but as a function of the horizontal range in yards.

Table 3: Sierra 0.308 caliber, 168 grain, MatchKing, $v_0 = 2650\,ft/s$

x_T	100	200	300	400	500	600	700	800
i	0.5116	0.5139	0.5162	0.5184	0.5207	0.5229	0.5252	0.5275
c	2.7288	2.7411	2.7533	2.7651	2.7773	2.7891	2.8013	2.8136
x_T	875	900	1000	1200	1500	1800	1968.5	2000
i	0.5291	0.5297	0.5320	0.5365	0.5433	0.5501	0.5539	0.5539
c	2.8221	2.8253	2.8376	2.8616	2.8979	2.9342	2.9544	2.9544

2.12. Maximum Range of Bullets

Theoretical calculation of the maximum range of bullets, at least for the non-military ballisticians and amateur riflemen, remains an open problem. [43].

The form coefficient estimated using Tzernozubovim's formula (2.6.8) and the property that the form coefficient remains constant for ranges

$$x_T \geq 1800 \cdot (d\,/\,0.0762) \qquad\qquad (2.12.1)$$

allow us to have an estimate on the maximum range of the trajectory of a bullet.

Theoretical-Experimental Method

To illustrate the method of calculating theoretically the maximum range of a bullet, let's consider example 10.2. The ballistics coefficient $c=3.1336$ of a 7.62mm Ball M2 Bullet (departure speed 840m/s) corresponds to ranges greater than or equal to 1,800 meters.

---

43. Hatcher, J. S, "Hatcher's Notebook", p. 541, 3d edition, Stackpole Books, 1966
Rinker, R.A., "Understanding Firearm Ballistics", p.265, 6th edition, Mulberry Publishing House, 2005

Using the PC program RPROJC.BAS (section 2.17), (departure speed:, 840m/s; temperature of air and projectile:, 15 degrees; atmospheric pressure, 750mm Hg; pressure of water vapors, 6.35mm Hg.; BC, 3.1336), for different values of the departure angle α_0, we find the following ranges x_T :

$\alpha_0 = 30°$	$x_T = 3499m$	$v_T = 108m/s$ $t_T = 27.1s$
$\alpha_0 = 31°$	$x_T = 3502m$	$v_T = 108m/s$ $t_T = 27.7s$
$\alpha_0 = 32°$	$x_T = 3503m$	$v_T = 109m/s$ $t_T = 28.3s$
$\alpha_0 = 33°$	$x_T = 3501m$	$v_T = 110m/s$ $t_T = 29.0s$

From the above data, we find that the maximum range, around $x_T = 3503m$, is obtained when the departure angle is approximately $\alpha_0 = 32°$.

The theoretical result needs to be verified with experimental shootings.

Since it is difficult for the shooter to observe the point of impact of the bullet on the ground or on the surface of water (though we have a calculated range as a guide to the possible point of impact), it is recommended to carry out the experiments in winter time, firing on the frozen rivers, or lakes, where it is not difficult to find the bullet trace..

Where to expect the point of impact in winter?

Let's assume that at the firing site, the temperature of air is (-5) degree Celsius, pressure is 760mmHg, humidity is 50%, - and pressure of water vapor is 1.63mmHg. The temperature of propellant is (-5 degree Celsius).

Using again the PC program RPROJC.BAS, we find:

$\alpha_0 = 30°$	$x_T = 3236m$	$v_T = 103m/s$ $t_T = 26.1s$
$\alpha_0 = 31°$	$x_T = 3238m$	$v_T = 104m/s$ $t_T = 26.6s$
$\alpha_0 = 32°$	$x_T = 3238m$	$v_T = 105m/s$ $t_T = 27.3s$
$\alpha_0 = 33°$	$x_T = 3236m$	$v_T = 105m/s$ $t_T = 27.8s$

We find that the bullet strikes the frozen surface at the range x_T = 3238m when the bullet is fired at an angle of α_0 = 32° .

Let's now assume that the observations show that the bullet falls at maximum range of 3,150 meters when it is fired with an angle α_0 = 32° .

Using the PC Program BCPROJ.BAS and the maximum, range found experimentally, we find that the ballistics coefficient is c = 3.2315.

Employing the obtained value of the ballistics coefficient, we are able to estimate the maximum range in whatever atmospheric conditions, in presence or absence of wind, etc.

For the normal conditions:

departure speed, 840m/s; temperature of air and temperature of projectile, 15 degrees; atmospheric pressure, 750mmHg.; pressure of water vapors, 6.35mmHg.; BC, 3.1336, using the PC Program RPROJC.BAS, we find that the maximum range is x_T = 3418m and is reached when the departure angle is around α_0 = 32° .

The terminal speed is v_0 = 107m/s. The impact energy of the bullet is E_K = 54.18J .

Note:

To estimate approximately the maximum range, the shooter can use the ballistics coefficient given by the manufacturer, though it is valid for short distances.

The maximum range estimated using the manufactures BC would normally be greater than the real range.

2.13. Equivalent Ballistics Coefficient

The ballistics coefficients, given by manufacturers of the firearms and projectiles, are valid only when they are used with the corresponding recommended G-function.

One of the problems of the exterior ballistics is the determination of the ballistics coefficient relative to a G-function, when it is known the ballistics coefficient of the same projectile relative to another G-function.

The analytical G-functions obtained in sections 1.4-1.6, allow us to find an "equivalent" ballistics coefficient related with any G-function.

Let's illustrate the method with the following example

The form coefficient of a 7.62mm Russian bullet, within the effective range of the bullet, fired with an initial speed of 735m/s is $i=0.56$. The corresponding ballistics coefficient is $c=4.116$. That value of the ballistics coefficient correspond to a range of 500 meters and the departure angle 0.432 degree, when the projectile is fired in the TSA atmosphere (temperature 15 degrees, pressure 750 mm, pressure of water vapors 6.35mmHg. at the altitude of 110 meters over the sea level).

The terminal speed at 500 meters is 373m/s.

Find the equivalent ballistics coefficient of the given projectile that corresponds to the G_8 function in ASM atmosphere.

(a) Consider the Siacci G-function, (1.1.6), and the analytical G_8 function, (1.4.13), for projectile speed greater than 256m/s, i.e., respectively:

$$K_D(v) = v/3 - 80, \quad G_8(v) = 0.14733 \cdot v - 29.085. \quad (2.13.1)$$

We denote c_S the Siacci ballistics coefficient ($c_S = 4.116$) and c_8 the ballistics coefficient related to G_8 function.

Consider that the projectile is launched in the ASM atmosphere. Using (2.2.10),

$$a_D = c \cdot \frac{p_0}{p_{0N}} \cdot h_0(y) \cdot G_D(v \cdot \sqrt{\tau_{0N}/\tau_0}),$$

we can write:

$$a_D = c_S \cdot \frac{p_0}{p_{0N}} \cdot h_0(y) \cdot (v - 80/3) \qquad (2.13.2)$$

and

$$a_D = c_8 \cdot \frac{p_0}{p_{0N}} \cdot h_0(y) \cdot (0.14733v - 29.085) \qquad (2.13.3)$$

Hence,

$$c_8 = \frac{v/3 - 80}{0.14733v - 29.085} \cdot c_S . \qquad (2.13.4)$$

The above formula shows that the ballistics coefficient depends on the speed of projectile. The average value of the ballistics coefficient in the given interval is

$$c_8 = c_S \int_{373}^{735}(v/3 - 80)dv \div \int_{373}^{735}(0.14733v - 29.085)dv = 4.117 \cdot \frac{37889.333}{19017.967} = 8.2003$$

The corresponding BC ins English units is

$$C = \frac{1.4222}{c} = \frac{1.4222}{8.2003} = 0.1734$$

Comments:

The results obtained in the following examples, 13.1 and example 13.2, show that:

- Within the effective ranges of small firearms, the BC converting method gives good results for G_1, G_7, G_2, functions and somehow for G_5 function of resistance.
- For long ranges, the BC converting method gives approximate results.
- The equivalent ballistics coefficient is an approximate value that can serve to compare the ballistics coefficients of projectiles of different manufacturers within the effective range.
- The discrepancies in the trajectories of flight obtained using the equivalent ballistics coefficients are due to the fact that the ballistics coefficient depends on the departure angle. A

fixed ballistics coefficient can not give accurate results especially for relatively long ranges.

We conclude that the best way to estimate the BC relative to a given G-function, especially for long distances, is the experimental method.

Example 13.1 Equivalent BC(s) of G_I M80 Ball Bullet

A G_I M80 ball bullet (mass 149 grain, FMJ boat tail) [44], launched horizontally at a speed of 838.2m/s, has a ballistics coefficient of $C_7 = 0.195 lb/in^2$. The terminal speed of the projectile at the range of 1,000 yards is 309.07m/s.

Find the ballistics coefficient of the given bullet that corresponds respectively to G_1, G_2, G_5, G_6, and G_7.

Solution:

BC relative to the G_1-function

Considering the relationship between the BC in English units and the corresponding BC in metric units,

$$C = \frac{1.4222}{c},$$

using the G_7 and G_1 functions given respectively by (1.4.12) and (1.4.5) and the method employed above, we find that the BC of the given bullet with respect to G_1 is

$$\frac{1}{C_1} = \frac{1}{C_7} \int_{3.0907}^{838.2} (0.150355v - 34.7319)dv + \int_{309.07}^{838.2} (0.312914v - 79.3876)dv = (\frac{1}{0.195}) \cdot (\frac{27259.185}{52971.639}) = 2.63897$$

Hence, we find that the BC related with the G_1 function is

$$C_1 = \frac{1}{2.63897} = 0.3789 lb./in^2 .$$

---

44. Schaefer, J.C. A, "Brief Discourse on Ballistics Coefficients", www.frfrogspad.com/extbal.htm (Web access November, 2009)

The ballistics coefficient expressed in metric units is

$$c_1 = \frac{1.4222}{C_1} = 3.7535 m^2 / kg .$$

Using McCoy's PC program MCTraj41.exe and the estimated ballistics coefficient, we find that the path of projectile for 1,000 yards is -407.1 inches.

In the following table there are presented the data obtained by Schaefer (first two columns) using respectively $C_7=0.195 lb/in^2$ and $C_1=0.393 lb/in^2$, and the results obtained using McCoy's PC program MCtraj41.exe (the last two columns of table 1), for $C_7=0.195 lb/in^2$ and $C_1=0.3789 lb/in^2$ (distances of 500 yards, 800 yards, and 1,000 yards).

Note:

According to Schaefer, the BC of the given bullet with respect to G_1-function of resistance is between $C_1 = 0.393$ and $C_1 = 0.395$.

Table 1: GI M80 ball bullet

Range Yard	$C_1 = 0.393 lb / in^2$ Path	$C_7 = 0.195 lb / in^2$ Path	$C_1 = 0.3789 lb /$ Path	$C_7 = 0.195 lb / in^2$ Path
500	-36.8	-37.0	-37.6	-37.1
800	-188.4	-191.7	-194.6	-192.2
1000	-391.8	-408.4	-407.1	-409.5

The BC for distances of 500 meters and 800 meters

The projectile speed at the distance of 500 meters and 800 meters are respectively 525m/s and 387m/s.

Using the same procedures we have used above, we obtain the following ballistics coefficients:

$$C_1 = 0.3854 lb./in^2 \text{ and } C_1 = 0.3788 lb./in^2 .$$

The BC relative to G_2, G_5, G_6, and G_8

In the same way as in the above calculations, using the G-functions, G_2, G_5, G_6, and G8, we find respectively the following corresponding ballistics coefficients:

$$C_2 = 0.1936 lb / in^2, \quad C_5 = 0.2427 lb / in^2,$$
$$C_6 = 0.2156 lb / in^2, \quad C_8 = 0.2098 lb / in^2.$$

In table 2 are displayed the data obtained using McCoy's PC program Mctraj41.exe. Table 2 shows that for large distances, the discrepancies are relatively big, especially the results obtained using G_6 and G_8 functions. Maybe, the errors are result of the fact that the ballistics coefficients are valid for the effective ranges of small arms.

Table 2.

Range Yard	$C_2=0.1936 lb/in^2$ Path	$C_5=0.2427 lb/in^2$ Path	$C_6=0.21 lb/in^2$ Path	$C_8=0.2098 lb/in^2$ Path
500	-36.7	-37.7	-38	-37.3
800	-192	-197.2	-202.5	-196.3
1000	-413	-416.6	-436.8	-422.7

Example 13.2 Equivalent BC(s) of 7.62mm M80 Ball, 24" Barrel

A 7.62mm M80 ball bullet (mass 148 grain, FMJBT) [45], launched horizontally at a speed of 870m/s, has a ballistics coefficient $C_1=0.404 lb/in^2$ (relative to G_1-function). The terminal speed of the projectile at the range of 1,000 m is 324m/s. In column 1 of table 3, there is given the path of the bullet (zeroed at 100m). The data in column 1 are from the range table of the given firearm and bullet ($C_1=0.404 lb/in^2$).

Find the ballistics coefficients of the given bullet that corresponds respectively to G_1, G_2, G_5, G_6, G_7, G_s (Siacci).

---

45.   http://www.hpbt.org/articles/308.doc

Solution:

Using the same procedure we used in Example 12.1, we find the following ballistics coefficients:

$C_2=0.2048lb/in^2$, $C_5=0.2585lb/in^2$,
$C_6=0.2266lb/in^2$, $C_7=0.2070lb/in^2$,
$C_8=0.2214lb/in^2$, (Siacci BC): $C_S=0.4476lb/in^2$.

The path of the bullet (zeroed at 100 meters) that is calculated using the McCoy's PC program Mctraj41.exe and the calculated ballistics coefficients is given in table 3.
The second column of table 3 shows the data given by the respective reference.

Table 3

Range Meters	C_1=0.404 Path	C_2=0.2048 Path	C_5=0.2266 Path	C_6=0.2266 Path	C_7=0.207 Path	C_8=0.2214 Path
300	-16.6	- 16.4	-16.6	-16.6	-16.6	-16.5
500	-69.8	-68.5	-69.7	-70.2	-69.3	-69.3
800	-262.7	259.4	-264.9	-272.4	-260.4	- 264.8
1000	-515.6	521.1	-524.3	-552	-518.1	- 533.7

As we see from table 3, the equivalent ballistics coefficients give approximate results especially for ranges less than 800 meters.
Note that a fixed BC does not give accurate results for large horizontal ranges, since the BC is a function of the departure angle.

2.14. PC Program to Calculate the Ballistics Coefficient

The PC program BCPROJ.BAS can be used to find the ballistics coefficient (in metric units),

$$c=\frac{id^2}{m}\cdot1000,\qquad(2.14.1)$$

that corresponds to any G-function using the experimental data obtained in whatever atmospheric conditions or in high-altitude shootings.

Using the BCPROJ.BAS, we can solve the main problem of exterior ballistics related with the determination of the ballistics coefficient that in general can be formulated:

- Find the ballistics coefficient of a projectile c when it is known the departure speed, departure angle, and the horizontal range are known.
- Find the ballistics coefficient of a projectile c when it is known the departure speed, and the drop of a projectile for a given horizontal range are known.
- Find the ballistics coefficient of a projectile when it is known the departure speed, the departure angle, and the coordinates of the firearm and the target are known.

For a given atmosphere (ASM, ICAO, TSA), the ballistics coefficient is a function of the departure angle, but it does not depend on the changes of the characteristics of air at the departure point of the projectile. In other words, for a given angle of departure, the ballistics coefficient does not change if the atmospheric characteristics of air at the shooting site are different from those accepted as standard.

The PC program determines the BC of the projectile, related with a given G-function of resistance, no matter if the projectile is launched in standard or non standard atmosphere.

The ballistics coefficient changes if the ballistics characteristics of the projectile (mass, caliber, form, propellant charge) change from those accepted as standard.

For the same projectile, the BC depends as well on the departure speed (length of firearm barrel, or ballistics characteristics of the propellant (the temperature of propellant, etc.).

Anyway, the BCPROJ.BAS calculates the BC of the fired projectile, no matter if the projectile has the standard characteristics or not.

We illustrate the use of BCPROJ.BAS through exercises.

Remarks:

- The guessed value of the BC we need to input in the BC.BAS can also be zero.
- The value of a ballistics coefficient "c" might slightly change if we input in the PC program COEFF.BAS another guessed BC, or another value of the "error.".
- The origin of the rectangular coordinative system is at the muzzle of the firearm no matter what is the altitude of firing site over the sea level. The y-coordinate of the firearm located at the firing site should always be zero. The y-coordinate of the target is zero, when the target is at the same altitude as the firearm; it is positive when the target is over the firing site and negative when the target is under the firing site.
- The program might enter in a cycle. In this case, the values of the BC presented in the second column, alternates between two approximate values. For the value of BC, we can choose one of those two displayed values. Stop the program. To get a value of the BC, we must again execute the program after changing the x-coordinate error.
- There might be cases where the program does not give an answer, and the message overflow does appear. The reason might be that the input data do not correspond to a real shooting; for example, the departure angle does not correspond to the practical range of shooting.
- The BC estimated for any range depends on the error in x-coordinate we input in the program. If, for example, we want an accuracy of 1cm (0.01m) in the vertical direction, we should input an error calculated using the formulas:

$$Error = \frac{0.01}{\tan(\alpha_T)}, \text{ or } Error = \frac{0.01}{\tan(\alpha_0)}, \tag{2.14.2}$$

Example 14.1 BC and the Horizontal Range

A 172-grain bullet, caliber 0.30-caliber M1 fired with a speed of 2700fps (823m/s), at an angle of 52 minutes (0.86666 degree),

hits the ground at the horizontal distance of 1,200 feet (1097.30m). [46].
Find the BC of the given bullet relative to G_1. The firing site is at
the sea level and the atmospheric conditions are those of the
ASM atmosphere.

Solution:

Using the PC program BCPROJ.BAS,
Input: ASM; G_1 function; Guess initial coefficient, 4; Projectile
speed, 823; Departure angle, 0.86666; Horizontal range, 1097;
Altitude of target, 0; Altitude of gun, 0; Error, 0.2; Integration step, 1;
temperature of air, 15; temperature of propellant, 15; atmospheric
pressure, 750; pressure of water vapors, 9.

Output: Ballistics coefficient relative to the G_1 function
$c=2.6551m^2/kg$.

In English units, the BC is

$$C = \frac{1.4222}{c} = \frac{1.4222}{2.6551} = 0.5356lb / in^2$$

Note:

Using the data given in *Hatcher's Notebook* (p.402) and the
BCPROJ.BAS, we find the following values of BC:

    Range, 800 yards: c = 2.5524; C = 0.5572
    Range, 500 yards: c = 2.5026; C = 0.5683
    Range, 300 yards: c = 2.4613; C = 0.5778

The range table in *Hatcher's Notebook* (p.402) is obtained using
the Siacci's solution method, the J(v) table, and a fixed BC =
0.2702.

---

46. Hatcher J. S., "Hatcher's Notebook", page 402, Stackpole Books, 1962.

Comments:

Since the data in *Hatcher's Notebook* are calculated theoretically, there might be differences in the results that can be obtained by testing fires. Anyway, the BC obtained using the BCPROJ.BAS represents the BCs that corresponds to the theoretically calculated data.

Exercise 14.2 BC and the Maximum Range

The maximum range of a 0.30-06 Ball M2 bullet, fired at an initial speed of 853.44m/s (2800 fps), is 3,200m (3,500 yards). Find the BC of the bullet related to the G_1 function considering that the departure angle is around 32 degrees. The shooting site is at the sea level and the atmospheric characteristics are those of the ASM atmosphere.

Answer:

$c = 5.8988; C = 1.581,$

Example 14.3

The range tables of Russian projectiles fired from small arms are constructed using the experiments performed at an altitude of 110 meters over the sea level considering the following characteristics of air (at the given altitude):

- Atmospheric pressure is 750mmHg.
- Air temperature is 15 degrees Celsius.
- Air density is, $1.206kg/m^3$.
- Air humidity, is 50%.

Use the PC program BC.BAS to find the ballistics coefficient of a 7.62mm projectile fired with a speed of 735m/s from a 7.62 mm Russian light machine gun "Degtyaryov" at an angle of 1.62 degree, if the horizontal range is 1,000 meters.

Solution:

Using the PC program BC.BAS

Input: Traditional Standard Atmosphere, TSA ,3; Siacci G-function; Guess initial coefficient, 4; Projectile speed, 735; Departure angle, 1.62; Horizontal range, 1000; Altitude of target, 0; Altitude of gun, 0; Error, 0.1; Integration step, 10.

Output: Ballistics coefficient is 4.4048.

Exercise 14.4

The range tables of Russian projectiles fired from small arms are constructed using experiments performed at an altitude of 110 meters over the sea level considering the following characteristics of air (at the given altitude):

- Atmospheric pressure is 750mmHg.
- Air temperature is 15 degrees Celsius.
- Air density, is $1.206kg/m^3$.
- Air humidity, is 50%.

Find the ballistics coefficient for a Russian 7.62mm bullet fired from a 7.62mm carbine "Semjonov" considering the following data:

At the firing site: initial speed, 735m/s; departure angle, 0.432 degree; horizontal range, 500m.

Answer:

BC = 4.135

Exercise 14.5 BC in High-Mountain Shooting

A projectile of the Russian cannon 122mm, Mod.1960 is fired at the altitude of 1,500 meters over the sea level.

The temperature, the pressure, and the water vapor pressure at the shooting site are respectively 5.508 degrees Celsius, 625.09mmHg, and 3.5mmHg. Consider that the temperature of the propellant is 15 degrees Celsius.

Note that the atmospheric conditions are TSA standard (at the sea level: the temperature is 15 degrees Celsius, the pressure is 750mmHg, and the pressure of water vapors is 6.35).

(a) Use the BCPROJ.BAS to find the BC of the projectile (relative to the Siacci G-function and relative to Russian G-function of the year 1943) considering the following test data:

Departure speed, 885m/s; horizontal range, 15,800 meters; departure angle 12.75 degrees.

(b) Is the BC found in (a) valid also for the flight of the same projectile at the sea level?

Answer:

(a) $BC_S = 0.2463$ (Siacci); $BC_{43} = 0.5084$. (b) yes.

Comment:

The BC found in (a) is also valid for the projectile fired at the sea level, at the same initial speed and departure angle.

Exercise 14.6

Find the ballistics coefficient (relative to the Siacci's G-function) of a Russian 122mm artillery gun, Mod. 1960, if a projectile fired with initial speed 885m/s at an angle of 13.68333 degrees, will hit the ground at a range of 15,000m. Consider the TSA atmosphere.

Answer:

$BC = 0.2502$.

Example 14.7 BC and the Projectile Drop

The drop of a Sierra NATO bullet caliber 0.30, 168 grain HPBT, fired horizontally, at the sea level, with initial speed 808m/s at the horizontal range of 550 meters is 3.2 meters.

Find the ballistics coefficient that corresponds to G_1, G_7, and G_8 functions, assuming that the projectile flight is in the ICAO atmosphere.

Solution:

Using the PC program BCPROJ.BAS.,

Input: G-function, G_1; Atmosphere, ICAO; Guess a BC, 3; Speed of projectile, 808; Departure angle, 0; x-coordinate of target, 550; y-coordinate of target, 0; y-coordinate of gun, 3.2; error, 0.5, number of steps, 1.

Output: Ballistics coefficient that corresponds to the G_1-function is $c=3.061$.s

In the same way, we find that the BC for the G_7-function, and the G_8-function, are respectively $c = 6.062$ and $c = 5.774$.

The corresponding ballistics coefficients in English unit, estimated using

$$C = \frac{1.4222}{c},$$

are respectively:

$$C = \frac{1.4222}{c} = \frac{1.4222}{3.061} = 0.4646 lb\,/\,in^2, \quad C = 0.2346 lb\,/\,in^2,$$

$$C = 0.2463 lb\,/\,in^2, \text{ or } C = \frac{1.4222}{c}\frac{d^2}{d_e^2}.$$

Exercise 14.8 BC and the Projectile Drop

The drop of a 5.56mm M855 (62 grain) bullet, fired horizontally, at the sea level, with initial speed of 945m/s at the horizontal range of 600 meters is -3.795 meters.

Find the ballistics coefficient that corresponds to G_7 function, assuming that the projectile flight is in the ASM atmosphere.

Answer:

(a) $c = 10.246$ (input error 0.8); $C = 0.1388$.

2.15. Ballistics Coefficient and the Bullet Drop

Nowadays, the Doppler radar measurement technique that is used to determine the elements of the projectile trajectories, is the most accurate method to create the range tables.

The radar measurements of the elements of the trajectories of bullets fired by small firearms give us the possibility to find the ballistics coefficients as a function of the horizontal range and the projectile drop.

The Doppler radar is also used to find the G-function of a given projectile in table form. Such G-functions are for example the G-functions of the Lapua bullets.

The traditional ballistics is not based on Doppler radar equipments, but still uses the traditional methods based on the standard G-functions (G_1, G_2, ... etc.) and the corresponding BC(s).

The ballistics coefficient of a projectile can be found using the PC program BCPROJ.BAS, the Siacci's solution method or other approximate methods.

The following example demonstrates the method we can use to find the ballistics coefficients.

Consider the Lapua Scenar GB528, 19.44 g (300 grain) bullet and the respective data in table 1 obtained by radar measurements. [47]. The Lapua Scenar GB528, 19.44 g bullet is fired horizontally at the sea level, in the ICAO atmosphere with an initial speed of 830 m/s.

---

[47]. Wikipedia Contributors, "External Ballistics", Wikipedia, The Free Encyclopedia, http://en.wikipedia.org/wiki/External_ballistics (accessed October 24, 2009).

Table 1: Radar Obtained Data: Lapua Scenar GB528 19.44 g Bullet Drop

Range	0	300	600	900	1200	1500	1800	2100
Speed	830	711	604	507	422	349	311	288
Time	0	0.392	0.851	1.394	2.044	2.828	3.748	4.752
Drop	0	- 0.715	- 3.203	- 8.146	- 16.57	- 30.04	- 50.72	- 80.53

In table 2, it is shown the ballistics coefficient relative to G_1-function (ICAO atmosphere),

$$c = \frac{id^2}{m} 1000,$$ (2.15.1)

as a function of the horizontal range and the projectile drop.

The ballistics coefficients presented in table 2 are obtained using the PC program BCPROJ.BAS. In table 2, there are presented as well the terminal speed, the projectile drop, and the time of flight obtained using the RPOINT.BAS (section 2.18) and the BC already found using BCPROJ.BAS.

In table 2 is also shown the ballistic coefficient in English units (in/lb):

$$C = \frac{1.4222}{c}$$ (2.15.2)

Table 2: The ballistic coefficient as a function of the horizontal range

Range [m]	300	600	900	1200	1500	1800	2100
c [m/kg]	1.9644	1.9019	1.8735	1.8655	1.8740	1.8929	1.9040
Speed	705	597	503	423	359	312	283
Time	0.39	0.853	1.396	2.045	2.823	3.743	4.776
Drop	- 0.715	- 3.203	- 8.148	-16.585	-30.105	-50.948	-81.18
C [in^2/lb]	0.7240	0.7478	0.7591	0.7624	0.7589	0.7513	0.7470

The use of both PC programs to obtain the ballistics coefficients and the firing data, presented in table 1 and table 2, is demonstrated below.

Range 600, BCPROJ.BAS
Input: Atmosphere? Press 2; G-Function? Press, 1; Y/N? Press Enter; Initial Coefficient? 2; Speed? 830; Angle? 0; x-coordinate? 600; y-coordinate of target? 0; y-coordinate of gun? 3.203 (note that the coordinate of gun is equal to the absolute value of drop, -3.203); Error? 0.2; Number of steps? 10; Temperature? 15; temperature of propellant 21.1; Pressure? 760; Pressure of water vapor? 0.

Output: BC is 1.9019.

Range 600, RPOINT.BAS:
Input: Atmosphere? Press 2; G-Function? Press, 1; y-coordinate of firearm? 0; departure speed? 830; Departure angle? 0; Temperature of air? 15; Temperature of propellant 21.1; Pressure? 760; Pressure of water vapor? 0; Projectile mass: 0.01944; Change in mass? 0; Range wind? 0; Crosswind? 0; x-coordinate of a point? 100; x-coordinate of a pont? 600; Ballistics coefficient? 1.9019; Step? 1.

Output: Second Point, 600m: Drop (y-coordinate) is -3.203m; Time is 0.853s; Speed is 597m/s.

Example 15.1 Projectile Drop and Departure Angle

(a) Find the departure angle for the Lapua bullet presented in the section above for the ranges of 300m, 600m . . . 2,100m and the ballistics coefficient function $c=f(\alpha_0)$.
(b) Find the departure angle that corresponds to 500 meters and the corresponding ballistics coefficient.

Solution:

Hereafter, we demonstrate the solution for 600 meters. We can use the property rigidity of trajectory to zero the trajectory at 600 meters (see *Exterior Ballistics of Small Arms*, section 1.9).

$$\alpha_0 = ar\tan(\bar{y} / x_T) = ar\tan(3.203 / 600) = 0.30586° .$$

The results obtained for all distances are shown in table 1 below. Table 1 represents the ballistics coefficient function in tabular form.

Table 2: The ballistic coefficient as a function of the horizontal range

Range [m]	300	600	900	1200	1500	1800	2100
α_0 [degree]	0.1366	0.3059	0.5187	0.7918	1.1498	1.6213	2.2138
c [m/kg]	1.9644	1.9019	1.8735	1.8655	1.8740	1.8929	1.9040
C [in²/lb]	0.7240	0.7478	0.7591	0.7624	0.7589	0.7513	0.7470

(b) Using table 2, by interpolation, we find α_0 = 0.2495° and
     c = 1.9227 (C = 0.7397).

QBasic PC Program BCPROJ.BAS
'Finding Ballistics Coefficient Automatically
'Ballistics Coefficient (BC): c=1000id^2/m
'The computed coefficient "ko" is reserved in the file c:/koef.dat
'---

'Control Example:
'Estimate the BC of a projectile corresponding to G2 function:
'Initial speed = 885m/s; departure angle = 6.2 degree; Range = 10000m
'Atmosphere: ASM

'SOLUTION PROCEDURE
'Input Data
'Cancel data File : Print y/n

'Input "G-Function; G1= 1, G2=2, G5=5, G6=6, G7=7, G8=8, Siacci = 10"; 2
'Input "Atmosphere: ASM = 1, ICAO = 2; Standard = 3"; 1

'INPUT "Guess Initial Coefficient =   "; 0.2
'INPUT "Departure Angle [Degree.Minutes]   "; 6.2
'INPUT " Initial speed [m/s]   "; 885
'INPUT "x-coordinate of Target [m] =   "; 10000
'INPUT "Y-coordinate of Target [m]   "; 0
'INPUT "y-coordinate of GUN   "; 0
'INPUT "Number of Steps n   "; 100
'INPUT "Error in x-coordinate   "; 0.5
'INPUT "Temperature at the sea level   "; 15
'INPUT "Pressure at the sea level   "; 750
'Input "Pressure of air Vapor at Sea level "; 9.906

'Output:
' Ballistics Coefficient BC   = 0.5308
'X-coordinate of Target/Range = 10000
'Departure angle   = 6.2
'Impact Speed   = 406

'---
'Functions, Subs

DECLARE SUB y1z1v1w1 (x, y, z, v, w, y1, z1, v1, w1, koef, E, F, D, TE, pa1, bb, ta1, Pr, pa)
DECLARE SUB InfHyres (koef, kk, dis, n, speed, gab, yy, vo, G, GA, GS, E, F, D, atm, TE, ta, pa, ea, pa1, Pr)
DECLARE SUB NPxyzvw (nk, x, x0, y, y0, z, z0, v, v0, w, w0, h, h0, k, l, r, Q)
DECLARE SUB NPkoef (k, l, r, Q, h, y1, z1, v1, w1)
DECLARE SUB menu (cog, cof, xf, yf, xfu, yfu, t$)
DECLARE SUB Rezervim (koef, kek, gab, x0)

DECLARE SUB KthimiKendit (kk)

'Variables

DIM m(4, 4), v(4)
rendi = 4
cog = 7: cof = 0
gab = gab
dkoef = .0001
menu cog, cof, 3, 10, 21, 70, "INPUT"
InfHyres koef, kk, dis, n, speed, gab, yy, vo, G, GA, GS, E, F, D, atm, TE, ta, pa,
ea, pa1, Pr
CLS
PRINT "First Angle="; kk
hap = n
KthimiKendit kk

'Solution
ff:
x0 = 0: y0 = kk: z0 = y0: v0 = vo: w0 = 0: xx = dis: h0 = hap
y0 = speed * COS(y0 * 3.141516954# / 180)
z0 = TAN(z0 * 3.141516954# / 180)
F:
FOR nk = 1 TO rendi
NPxyzvw nk, x, x0, y, y0, z, z0, v, v0, w, w0, h, h0, k, l, r, Q
y1z1v1w1 x, y, z, v, w, y1, z1, v1, w1, koef, E, F, D, TE, pa1, bb, ta1, Pr, pa
NPkoef k, l, r, Q, h, y1, z1, v1, w1
m(nk, 1) = k: m(nk, 2) = l
m(nk, 3) = r: m(nk, 4) = Q
NEXT nk
FOR i = 1 TO rendi
v(i) = 1 / 6 * (m(1, i) + 2 * m(2, i) + 2 * m(3, i) + m(4, i))
NEXT i
'New Point
x0 = x0 + h: y0 = y0 + v(1): z0 = z0 + v(2)
v0 = v0 + v(3): w0 = w0 + v(4)

IF ABS(x0 - xx) <= .01 AND v0 <= (yy + gab * TAN(kk * 3.1415 / 180)) AND
v0 >= (yy + (-1 * gab) * TAN(kk * 3.1415 / 180)) AND ABS(x0 - xx) <= .01
THEN
'Display results
kek = gab * z0
impact = 180 * ATN(z0) / 3.141592654#
CLS
LOCATE 7, 21: PRINT "Ballistics Coefficient BC = "; koef
LOCATE 8, 21: PRINT "Range   = "; x0
LOCATE 9, 21: PRINT "Departure Angle = "; kk

```
LOCATE 10, 21: PRINT "Impact Speed  = "; y0
LOCATE 11, 21: PRINT "Time of Flight = "; w0
LOCATE 12, 21: PRINT "Impact Angle  = "; impact
LOCATE 13, 21: PRINT "Error in y-coordinate = "; ABS(kek)
LOCATE 14, 21: PRINT "Error in x-coordinate = "; ABS(gab)

PLAY "a8a16b8a8"
Rezervim koef, kek, gab, x0
dis = dis + 200
IF dis > disf THEN PRINT "END": INPUT b: GOTO fundi:
hap = n
READ kk
PRINT "New Angle ="; kk
KthimiKendit kk
GOTO ff:
fundi:
END
END IF
dkoef = .0001
IF ABS(x0 - xx) <= .01 AND v0 > (gab * TAN(kk * 3.1415 / 180)) THEN
koef = koef + dkoef
PRINT v0, koef
GOTO ff:
END IF
IF ABS(x0 - xx) <= .01 AND v0 < ((-1 * gab) * TAN(kk * 3.1415 / 180)) THEN
koef = koef - dkoef
PRINT v0, koef
GOTO ff:
END IF
GOTO F:
END

SUB InfHyres (koef, kk, dis, n, speed, gab, yy, vo, G, GA, GS, E, F, D, atm, TE,
ta, pa, ea, pa1, Pr)
CLS
LOCATE 4, 12: INPUT "Atmosphere: ASM = 1, ICAO = 2; Traditional Standard
= 3"; atmCLS
IF atm = 1 THEN GOTO 100:
IF atm = 2 THEN GOTO 200:
IF atm = 3 THEN GOTO 300:
100 LOCATE 4, 13: INPUT "G-Function; G1= 1, G2=2, G5=5, G6=6, G7=7,
G8=8, Ingalls= 10"; G
TE = 289.6: Pr = 750
IF G = 1 THEN E = .312914: F = 79.3976: D = .000100347#
IF G = 2 THEN E = .1413: F = 29.9097: D = .00009116233#
IF G = 5 THEN E = .19734: F = 49.0806: D = .0000743571#
IF G = 6 THEN E = .140533: F = 23.6633: D = .00010334#
```

```
IF G = 7 THEN E = .150355: F = 34.7319: D = .000056648#
IF G = 8 THEN E = .14733: F = 29.085: D = .000099366#
IF G = 10 THEN E = .32072: F = 82.6909: D = .000108774#
CLS
GOTO 400:

200 LOCATE 4, 13: INPUT "G-Function; G1= 1, G2=2, G5=5, G6=6, G7=7,
G8=8, Ingalls = 10"; GA
TE = 288.15: Pr = 760
IF GA = 1 THEN E = .31574: F = 78.6769: D = .00010584#
IF GA = 2 THEN E = .143353: F = 30.2415: D = .00009287#
IF GA = 5 THEN E = .200207: F = 49.625: D = .000075244#
IF GA = 6 THEN E = .142352: F = 23.6937: D = .000105244#
IF GA = 7 THEN E = .152593: F = 35.1717: D = .000057679#
IF GA = 8 THEN E = .149441: F = 29.379: D = .000101154#
IF GA = 10 THEN E = .325383: F = 83.6082: D = .00010724#
CLS
GOTO 400:

300 LOCATE 4, 13: INPUT "G-Function; Siacci = 1, Mayevski = 2, Ingalls = 3";
GS
TE = 289.08: Pr = 750
IF GS = 1 THEN E = .3333333#: F = 80: D = .000212
IF GS = 2 THEN E = .320243: F = 81.3721: D = .00010807#
IF GS = 3 THEN E = .200207: F = 49.625: D = .000075244#
CLS
GOTO 400:

400 LOCATE 5, 12: INPUT "Would You Like to Cancel the data File [Y/N]"; y$
    IF y$ = "y" OR y$ = "Y" THEN KILL "c:\koef.dat"
CLS

LOCATE 5, 12: INPUT "Guess Initial Coefficient  = "; koef
LOCATE 7, 12: INPUT "Projectile Speed         = "; speed
LOCATE 8, 12: INPUT "Departure Angle          = "; kk
LOCATE 9, 12: INPUT "x- coordinate of target   = "; dis
LOCATE 10, 12: INPUT "Y - coordinate of Target = "; yy
LOCATE 11, 12: INPUT "y - coordinate of the gun = "; vo
LOCATE 13, 12: INPUT "Error in x-coordinate    = "; gab
LOCATE 14, 12: PRINT "Number of Steps when range Ends with  0   is [10]"
LOCATE 15, 12: PRINT "Number of Steps when range Ends with 00   is [100]"
LOCATE 16, 12: PRINT "Number of Steps range Ends with a NON ZERO
Number is [1]"
LOCATE 17, 12: INPUT "Number of Steps          = "; n
LOCATE 19, 13: INPUT "Temperature of Air [C] at sea level     = "; ta
LOCATE 20, 13: INPUT "Atmospheric Pressure [mm] at sea level  = "; pa
LOCATE 21, 13: INPUT "Pressure of Water Vapor [mm] at sea level ="; ea
```

```
ta = ta + 273.15
pa1 = ta / (1 - .3785 * ea / pa)
CLS
END SUB

SUB KthimiKendit (kk)
kk = kk
END SUB

SUB menu (cog, cof, xf, yf, xfu, yfu, t$)
COLOR cog, cof
LOCATE xf - 1, yf: PRINT t$
LOCATE xf, yf: PRINT "É" + STRING$(yfu - yf, 205) + "»";
FOR i = xf + 1 TO xfu
LOCATE i, yf: PRINT "º" + SPACE$(yfu - yf) + "º";
NEXT
LOCATE xfu + 1, yf: PRINT "È" + STRING$(yfu - yf, 205) + "¼";
END SUB

SUB NPkoef (k, l, r, Q, h, y1, z1, v1, w1)
k = h * y1: l = h * z1
r = h * v1: Q = h * w1
END SUB

SUB NPxyzvw (nk, x, x0, y, y0, z, z0, v, v0, w, w0, h, h0, k, l, r, Q)
IF nk = 1 THEN
x = x0: y = y0: z = z0
v = v0: w = w0: h = h0
GOTO fund:
END IF
IF nk = 2 OR nk = 3 THEN
x = x0 + (.5 * h): y = y0 + (.5 * k)
z = z0 + (.5 * l): v = v0 + (.5 * r)
w = w0 + (.5 * Q)
GOTO fund:
END IF
IF nk = 4 THEN
x = x0 + h: y = y0 + k: z = z0 + l
v = v0 + r: w = w0 + Q
END IF
fund:
END SUB

SUB Rezervim (koef, kek, gab, x0)
OPEN "c:\koef.dat" FOR APPEND AS #1
PRINT #1, koef, kek, gab, x0
CLOSE #1
```

END SUB

```
SUB y1z1v1w1 (x, y, z, v, w, y1, z1, v1, w1, koef, E, F, D, TE, pa1, bb, ta1, Pr,
pa)
ta1 = (TE / pa1) ^ .5
bb = y * SQR(1 + z ^ 2)
IF (bb * ta1) > 256! THEN
y1 = -1 * koef * (pa / Pr) * ((TE - .006328 * v) / TE) ^ 4.4 * (ta1 * E * bb - F) /
(bb)
ELSE
y1 = -1 * koef * (pa / Pr) * ((TE - .006328 * v) / TE) ^ 4.4 * D * (ta1 * bb) ^ 2 /
bb
END IF
z1 = -9.80665 / y ^ 2
v1 = z
w1 = 1 / y
END SUB
```

2.16. Projectile Trajectory Universal PC Programs

Hereafter there are presented three universal PC programs obtained using the system of differential equations (2.5.1):

- RPROJC. BAS: Calculates the range of projectile
- RPOINT.BAS: Calculates the projectile drop, elements of trajectory for inclined fire.
- APROJ.BAS: Calculates the departure angle of a projectile.

The PC programs can be used to solve any exterior ballistics problem for the projectile flight in any atmosphere. They are based on the known G-functions of resistance that are in use in modern ballistics.

The PC programs can be used as well for the projectiles flying in ASM atmosphere, or ICAO atmosphere, as well as to solve the exterior ballistics related with the Siacci, Mayevski, and G43 functions in TSA atmosphere.

The PC programs presented in this chapter can also be employed for projectiles of firearms that have no standard characteristics, and/or (and) when the ballistics characteristics of projectile and propellant charge change from their standard values.

All three PC programs consider as well the influence of wind in the trajectory of the flaight.

The programs do not include methods that are used to calculate the influence of the projectile spin, the Coriolis force, etc., in the projectile flight. They can be estimated using empirical formulas, or the data given in range tables and can be included in the PC programs.

Using the PC programs presented in the following sections, together with the PC program BCPROJ.BAS shown in section 2.15, we are able to solve a large variety of exterior ballistics problems encountered in the theory and practice of the flight of projectiles fired by non-reactive firearms.

Though all three programs use a constant ballistics coefficient that is related with a given G-function of resistance, they can be modified to be used as well when the ballistics coefficient depends on the departure angle, $c=f(\alpha_0)$, or on the projectile speed. The modification can be made using the PC programs ACA122.BAS and RAN122.Bas, presented in section 2.4.

Remarks on the Construction of PC Programs

The PC programs presented hereafter are constructed by modifying and updating the PC programs that we have already used in *"Exterior Ballistics with Applications"* and are present and used as well in the *"Exterior Ballistics of Small Arms"*.

2.17. PC Program to Calculate the Horizontal Range

The PC Program RPROJC.BAS presented at the end of this section, can be used to find the horizontal range and the elements of the trajectory for a given x-coordinate, when the departure speed, departure angle, and the ballistics coefficient of the projectile are known.

Though the RPROJC. BAS uses a fixed ballistics coefficient, it can be easily modified to be used when the ballistics coefficient is given (known) as a function of projectile speed, or as a function of the departure angle.

The following examples illustrate the use of RPROJC. BAS to find the projectile range when it is known the departure speed, firing angle, and the ballistics coefficient of projectile are known.

Note that when the projectile mass is standard, we input the projectile mass 1 and a change in the projectile mass of zero. Those values can be input in all PC programs no matter if we know or not the projectile mass.

Example 17.1 Sea-Level Fire

A Russian 122mm projectile is fired from the field cannon Mod 1960 with a speed of 885m/s at an angle of 6.2 degree.

Find the range considering the Siacci G-function for the TSA atmosphere if the firing site is at the sea level.

At sea level, the temperature of dry air and the atmospheric pressure are respectively 15.9 degrees Celsius and 750mmHg.

The ballistics coefficient is 0.2389. The temperature of the propellant charge is 15 degrees Celsius.

Assume that there is no wind at the firing site. The humidity is 50%. The pressure of water vapors that corresponds to humidity 50% and temperature of 15.9 degree is 6.815 mm Hg.

Solution:

Input: Atmosphere TSA: 3; Siacci G-function: 1; Departure speed: 885; Departure angle: 6.2; Temperature of air at firing site: 15.9; Temperature of propellant: 15; Atmospheric pressure at firing site: 750; Pressure of air vapors at firing site: 6.815; Projectile mass: 1; Change in projectile mass: 0; Range wind: 0; Cross wind: 0; Ballistics coefficient: 0.2389; X-coordinate of a point: 0; Integration step: 1.

Output: Range: 10010m; Time of flight: 16.75s; Terminal speed: 420m/s; Terminal angle: -10.1819 degree; Trajectory vertex: (5638, 347).

Example 17.2 Mountain Fire

A Russian 122mm projectile is fired from the field cannon Mod 1960 with a speed of 885m/s at an angle of 6.2 degrees. Find the range considering the Siacci G-function for the TSA atmosphere if the firing site is 1,500 meters over the sea level.

At the sea level, the temperature of dry air and the atmospheric pressure are respectively 15.9 degrees Celsius, and 750mmHg. The ballistics coefficient is 0.2389. The temperature of the propellant charge is 15 degrees Celsius. Assume that there is no wind at the firing site.

Find as well the elements of the projectile trajectory at the point with abscissa of 10,000 meters.

Solution:

To use the program, we need to find the temperature (in degree Celsius), the pressure (mmHg.), and the pressure of air vapor (mmHg.) at the firing site.

The temperature of dry air at the firing site is

$$T = T_0 - 0.006328y = (15.9 + 273.15) - 0.006328(1500) = 279.558 \ ^\circ K.$$

The temperature of dry air in degree Celsius is

$$T_c = T - 273.15 = 279.558 - 273.15 = 6.408 \ ^\circ C.$$

The atmospheric pressure at the firing site is

$$p = p_0 h_0(y) = p_0 \left(\frac{T_0 - 0.006328y}{T_0}\right)^{5.4} = 750\left(\frac{278.59}{289.05}\right)^{5.4} = 626.26 mmHg.$$

Using table 1 section 2.2, we find that the pressure of air vapors at the firing site (temperature of 6.408 degrees Celsius) is 3.645 mmHg.

Using PC RPROJC.BAS
Input: Atmosphere TSA: 3; Siacci G-function: 1; Departure speed: 885; Departure angle: 6.2; Temperature of air at firing site: 6.408; Temperature of propellant: 15; Atmospheric pressure at firing site: 626.26; Pressure of air vapors at firing site: 3.645; Range wind: 0; Cross wind: 0; Projectile mass: 1; Change in projectile mass: 0; Ballistics coefficient: 0.2389; X-coordinate of a point: 10,000; Integration step: 1

Output: Range: 10,624m; Time of flight: 17.08s; Terminal speed: 451m/s; Terminal angle: -9.712 degree; Trajectory vertex: (5919, 359.53).

The elements of trajectory at the point with abscissa x = 10,000 are:

Y-coordinate: 97.38m; time of flight: 15.7s; speed: 469m/s; angle: -8.054.

Note:

From the range table of the respective projectile for the altitude of 1,500 meters over the sea level, we obtain the following elements of the trajectory (for the data given above):

range: 10640m; time: 17s; speed: 448m/s; angle: - 9.7; altitude of vertex point: 354m.

Example 17.3 Shooting in Low Temperatures

A Russian 122mm projectile is fired from the field cannon Mod 1960 with a speed of 885m/s at an angle of 6.2 degree.

Find the range and the other elements of the trajectory considering the Siacci G-function for the TSA atmosphere:

At the firing site, the temperature of dry air and the atmospheric pressure are respectively -5 degrees Celsius, and 750mmHg. The ballistics coefficient is 0.2389. The temperature of the propellant charge is 15 degrees Celsius. (The projectiles were kept in a storage with a temperature of 15 degrees Celsius.

Assume that there is no wind at the firing site. The humidity is 50%. The pressure of water vapors that corresponds to humidity 50% and temperature of -5 degrees is 1.633mmHg.

Solution:

Input: Atmosphere TSA: 3; Siacci G-function: 1; Departure speed: 885; Departure angle: 6.2; Temperature of air at firing site: -5; Temperature of propellant: 15; Atmospheric pressure at firing site: 750; Pressure of air vapors at firing site: 1.633; Projectile mass: 1; Change in projectile mass: 0. Range wind: 0; Cross wind: 0; Ballistics coefficient: 0.2389; X-coordinate of a point: 0; Integration step: 1.

Output: Range: 9781m; Time of flight: 16.64s; Terminal speed: 406m/s; Terminal angle: -10.4104 degree; Trajectory vertex: (5537, 342.5).

Example 17.4 Bullet 5.56mm, M855 Shooting

A NATO 5.56×45mm bullet is fired from an M16 Rifle with a speed of 940m/s at an angle 0.34876 degree.

The ballistics coefficient related to the G_7-function of resistance is $c=9.4436m^3/kg$ ($C=0.1506lb/in^2$). The temperature of the propellant charge is 21.1 degrees Celsius. Assume that there is no wind at the firing site. The humidity is 0%.

(a) Find the range and all the other elements at the impact point considering G_7-function for the ICAO atmosphere if the temperature and pressure are standard (respectively 15 degrees Celsius and 760mm).

(b) What will be the range if we use the G_7-function that corresponds to the ASM atmosphere?

Solution:

(a) **Input**: Atmosphere ICAO: 2; G_7-function: 7; Departure speed: 940; Departure angle: 0.34876; Temperature of air at firing site: 15; Temperature of propellant: 21.1; Atmospheric pressure at firing site: 760; Pressure of air vapors at firing site: 0; Projectile mass: 1; Change in projectile mass: 0; Range wind: 0; Cross wind: 0; Ballistics coefficient: 9.4436; x-coordinate of a point: 0; Integration step: 1.

Output: Range: 600m; Time of flight: 0.98s; Terminal speed: 402m/s; Terminal angle: - **0.6147** degree; Trajectory vertex: (346, 1.2);

(b) We obtain identical results when we use the G_7-function related to the ASM atmosphere:

Input: Atmosphere ASM: 1; G_7-function: 7; Departure speed: 940; Departure angle: 0.34876; Temperature of air at firing site: 15; Temperature of propellant: 21.1; Atmospheric pressure at firing site: 760; Pressure of air vapors at firing site: 0; Projectile mass: 1; Change in projectile mass: 0; Range wind: 0; Cross wind: 0;

Ballistics coefficient: 9.4436; x-coordinate of a point: 0; Integration step: 1.

Output: Range: 600m; Time of flight: 0.98s; Terminal speed: 402m/s; Terminal angle: - **0.616** degree; Trajectory vertex: (346, 1.2).

Example 17.5 Bullet 5.56mm, M855, Mountain Shooting

A NATO 5.56×45mm bullet is fired from an M16 Rifle in a mountain site at an altitude of 700 meters. The initial speed of the bullet is 940m/s, while the departure angle is 0.34876 degree. The ballistics coefficient related to G_7-function of resistance is $c=9.4436m^3/kg$ ($C=0.1506lb/in^2$).

The temperature of the propellant charge at the firing site is 21.1 degrees Celsius. Assume that at the firing site there is no wind. The humidity is 0%.

(a) Find the range and all the other elements at the impact point considering G_7-function for the ICAO atmosphere if the temperature and pressure at the sea level are standard (respectively 15 degrees Celsius and 760mm).
(b) What is the altitude of projectile at the point with abscissa 600m?

Solution:

Since we do not have the atmospheric data for the shooting site, we can estimate the temperature and pressure using the respective data at the sea level and the known laws of temperature and pressure with the altitude over the sea level.

The temperature of the dry air at the firing site is

$$T = T_0 - 0.006328y = (15 + 273.15) - 0.006328(700) = 283.72 \, ^\circ K.$$

The temperature of dry air in degree Celsius is

$$T_c = T - 273.15 = 283.72 - 273.15 = 10.57 \, ^\circ C.$$

The atmospheric pressure at the firing site is

$$p = p_0 h_0(y) = p_0 (\frac{T_0 - 0.006328y}{T_0})^{5.4} = 760(\frac{283.72}{288.15})^{5.4} = 709.92 mmHg$$

Use the PC RPROJC.BAS

(a) **Input**: Atmosphere ICAO: 2; G_7-function: 7; Departure angle: 0.34876; Departure speed: 940; Temperature of air at firing site: 10.57; Temperature of propellant: 21.1; Atmospheric pressure at firing site: 709.92; Pressure of air vapors at firing site: 0; Projectile mass: 1; Change in projectile mass: 0; Range wind: 0; Cross wind: 0; Ballistics coefficient: 9.4436; x-coordinate of a point: 600; Integration step: 1.

Output: Range: 615m; Time of flight: 0.99s; Terminal speed: 414m/s; Terminal angle: -0.6037 degree; Trajectory vertex: (353m, 1.22m).

(b) At the point with abscissa x = 600m: projectile altitude over the firing site is 0.154m.; Time of flight is 0.96s; projectile speed is 423m/s; terminal angle is - 0.5556.

Note:

From the example, we see that the point of impact at 600 meters is 15.4cm over the center of the target. Even if we do not change the sight, the bullet will hit the target 15.4 centimeters over the center of the target.

Exercise 17.6 Bullet 5.56mm M855, Mountain Shooting

The NATO 5.56×45mm bullet of the example 5.5 is fired at the same site as in preceding example (altitude 700 meters), from an M16 Rifle. The initial speed of the bullet is 940m/s, while the departure angle is 0.34876 degree. The ballistics coefficient related to the G_7-function of resistance is $c=9.4436 m^3/kg$ ($C=0.1506 lb/in^2$). The temperature of the propellant charge at the firing site is the same as the temperature of air at the site.

Assume that there is no wind. The humidity is 0%.

(a) Find the range and all the other elements of the trajectory at the impact point considering G_7 -function for the ICAO atmosphere if the temperature and pressure at the sea level are standard (respectively 15 degrees Celsius and 760mm).

(b) What is the altitude of projectile at the point with abscissa 600m?

Answer:

(a) Range, 604m; time, 0.98s; speed, 413m/s; impact angle, -0.6008 degree; vertex (349m, 1.19m).

(b) At the point with abscissa 600m,: altitude of projectile is 0.04m.

Exercise 17.7 Bullet 5.56mm M855, Winter Shooting

A NATO 5.56×45*mm* bullet of the example 5.5 is fired at the sea level from an M16 Rifle. The initial speed of the bullet is 940m/s, while the departure angle is 0.34876 degree. The ballistics coefficient relative to G_7-function of resistance is $c=9.4436m^3/kg$ ($C=0.1506lb/in^2$).

The temperature of the propellant charge at the firing site is the same as the temperature of air at the site. Assume that there is no wind. The humidity is 70%.

(a) Find the range and all the other elements at the impact point considering G_7 -function for the ICAO atmosphere if the temperature and pressure at the sea level are (-5) degrees Celsius and 760mm.

(b) What is the range if the shooting will take place at a site that is 500 meters over the sea level (calculated temperature is -8.164 degrees Celsius, and calculated pressure is 712.82mm Hg)?

Answer:

(a) Range, 560m; time, 0.95s; terminal speed, 388m/s; terminal angle, 0.6166.

(b) Range 570m; time, 0.96s; terminal speed, 399m/s; terminal angle, 0.6031.

Exercise 17.8 Flight of Projectile in Free Space

Use the data of example 17.7 to find the range of the NATO 5.56×45mm bullet if the shooting takes place in the free space (drag force is zero, and atmospheric conditions do not influence the flight).

Answer:

Using the PC program Rprojc.Bas, we obtain:

Range, 1097m; time, 1.17; terminal speed, 940m/s; terminal angle, -0.34876; vertex (557m, 2m).

Note:

The ballistics coefficient of any in free space is zero. We find the same results using the formulas that describe the trajectory of flight in free space.

Example 17.9 BC as a Function of the Departure Angle

The ballistics coefficient estimated experimentally is a function of the departure angle α_0, i.e. $c = f(\alpha_0)$.

The range tables of the Russian 122mm-caliber field cannon caliber 122mm, for the projectile fired at the initial speed 885m/s are obtained using the following table of the ballistics coefficients related with the G-function of the year 1943: [48]

---

48. "Range Tables of 122mm cannon Model 1960 ", p. 117, Ministry of Defense of Albania, 1967
Note: As a matter of fact the departure angle presented in the table is the elevation angle. For simplicity, it is not considered the jump -2minutes. Actually, the ballistics coefficient 0.514 corresponds to the departure angle 14.9667 degree.

Table 1 The BC of 122mm projectile, departure speed 885m/s.

Elevation, E_0	≤5°	15°	25°	35°	45°
BC, $c = f(E_0)$	0.496	0.514	0.516	0.519	0.521

The BC displayed in the table above is a function of the elevation angle. The jump is -2 minutes.

To produce the range table of the respective projectile in standard atmosphere, as well as the range tables in whatever atmospheric conditions or mountain firing, we can use the PC program Rprojc.Bas.

Use the RPROJC.BAS to find the range of the given projectile if the departure angle is 8 degrees.

Solution:

The elevation angle that corresponds to the departure angle of 8 degrees is 8 degrees and 2 minutes, i.e., $E_0 = 8.03333°$.

From table 1, by interpolation, we find that the ballistics coefficient of the given projectile that corresponds to the departure angle of 8 degrees is $c = 0.50146$.

The ballistics coefficient, $c = 0.5014$ we obtain from table 1 if we consider that the departure angle equal to the elevation angle (jump angle 0), is slightly different.

Input: Atmosphere TSA: 3; G_{43}-function: 3; Departure speed: 885; Departure angle: 8; Temperature of air at firing site: 15; Temperature of propellant: 15; Atmospheric pressure at firing site: 750; Pressure of air vapors at firing site: 6.35; Projectile mass: 1; Change in projectile mass: 0; Range wind: 0; Cross wind: 0; Ballistics coefficient: 0.50146; x-coordinate of a point: 0; Integration step: 1.

Output: Range: 11523m; Time of flight: 20.83s; Terminal speed: 372m/s; Terminal angle: -14.224degree; Trajectory vertex: (6616m, 539m).

Note:

The firing table of the given projectile shows that the range is 11,549m.

The PC program RPROJC, Range of Projectile
'FIND : Range,and other Elements of te Trajectory, etc.
'GIVEN:Departure Speed, Departure Angle,Ballistics Coefficient
'---
' DATA
'Input: Initial y-coordinate =0, departure Angle; 13.3, departure speed
,885
'Temperature of Air, 15, temperature of propellant, 15;
'Pressure = 750, Pressure of Air vapor = 6.35, Projectile mass, 1;
'Change in Projectile mass = 0.
'Results: Range = 15016, Error in y-coordinate,0.146, Time of Flight =
15.46s,
'Terminal Speed = 318m/s, Terminal Angle = -25.8587 Degree
'Cross wind deflection, 144m; vertex (8801, 1256)
'---

'Functions & Subs.
DECLARE SUB y1z1v1w1 (x, y, z, v, w, y1, z1, v1, w1, koef, pa1, wind,
ys, yy, pa, ta1, TE, E, F, D, Pr)
DECLARE SUB InfHyres (x0, y0, z0, v0, w0, a, h0, ta, pa, ea, tp, ta1,
pa1, xx1, voo, vol, wind, koef, cw, vv, E, F, D, Pr, TE, m, dm, atm, G,
GA, GS, Tc)
DECLARE SUB NPxyzvw (nk, x, x0, y, y0, z, z0, v, v0, w, w0, h, h0, k,
L, r, q)
DECLARE SUB NPkoef (k, L, r, q, h, y1, z1, v1, w1)
DECLARE SUB menu (cog, cof, xf, yf, xfu, yfu, t$)
DECLARE SUB c (koef, m, dm, BC, a)

'Variables
DIM m(4, 4), v(4)
rendi = 4
cog = 7: cof = 0
'Zgjidhja
CLS

fillimi:
menu cog, cof, 3, 10, 21, 70, "INITIAL DATA"
InfHyres x0, y0, z0, v0, w0, a, h0, ta, pa, ea, tp, ta1, pa1, xx1, voo, vol,
wind, koef, cw, vv, E, F, D, Pr, TE, m, dm, atm, G, GA, GS, Tc
c koef, m, dm, BC, a

F:

```
FOR nk = 1 TO rendi
NPxyzvw nk, x, x0, y, y0, z, z0, v, v0, w, w0, h, h0, k, L, r, q
y1z1v1w1 x, y, z, v, w, y1, z1, v1, w1, koef, pa1, wind, ys, yy, pa, ta1,
TE, E, F, D, Pr
NPkoef k, L, r, q, h, y1, z1, v1, w1
m(nk, 1) = k: m(nk, 2) = L
m(nk, 3) = r: m(nk, 4) = q
NEXT nk

'Calculation

FOR i = 1 TO rendi
v(i) = 1 / 6 * (m(1, i) + 2 * m(2, i) + 2 * m(3, i) + m(4, i))
NEXT i

'New Data
x0 = x0 + h: y0 = y0 + v(1): z0 = z0 + v(2)
v0 = v0 + v(3): w0 = w0 + v(4)
IF ABS(z0) < .0001 THEN
ymax = v0
xmax = x0 + wind * w0
END IF

xxc = x0 + wind * w0
IF (xxc - xx1) <= .001 THEN
xc = xxc
yc = v0
Tc = w0
ac = (180 / 3.141592654#) * ATN(z0)
vc = y0 / COS(ATN(z0))
END IF

IF x0 > 10 AND v0 <= .005 THEN
'Display Resultst
menu cog, cof, 6, 20, 22, 72, "RESULTS:"
LOCATE 7, 25: PRINT "Horizontal Range [m]    = "; INT((x0 + w0 *
wind - v0 / z0) * 100 + .5) / 100
LOCATE 8, 25: PRINT "Coresponding y-Coord [m] = "; (v0 - v0)
LOCATE 9, 25: PRINT "Departure Angle [Deg.]   = "; INT((a) * 10000
+ .5) / 10000
LOCATE 10, 25: PRINT "Time of Flight [s]    = "; INT((w0) * 100 +
.5) / 100
```

```
LOCATE 11, 25: PRINT "Terminal Speed [m/s]    = "; INT((y0 * (1 + z0
^ 2) ^ .5) + .5)
LOCATE 12, 25: PRINT "Terminal Angle [Deg.]   = "; INT((ATN(z0) *
180 / 3.141593) * 10000 + .5) / 10000
LOCATE 13, 25: PRINT "Cross-Wind Deflection   = "; INT((cw * (w0 -
x0 / (voo * COS(a * 3.14159265# / 180)))) * 1000 + .5) / 1000
LOCATE 14, 25: PRINT "Trajectory Vertex [m]   = "; "("; INT((xmax) *
10 + .5) / 10; ","; INT((ymax) * 100 + .5) / 100; ")"
LOCATE 16, 25: PRINT "Point on Trajectory [m]   = "; "("; INT((xc) +
.5); ","; INT((yc) * 1000 + .5) / 1000; ")"
LOCATE 17, 25: PRINT "Time [s]                = "; INT((Tc) * 100 + .5) /
100
LOCATE 18, 25: PRINT "Corresponding Speed [m/s] = "; INT((vc) + .5)
LOCATE 19, 25: PRINT "Corresponding Angle [Deg] = "; INT((ac) *
10000 + .5) / 10000
LOCATE 20, 25: PRINT "Cross-Wind Deflection   = "; INT((cw * (Tc -
xc / (voo * COS(a * 3.14159265# / 180)))) * 1000 + .5) / 1000
LOCATE 22, 25: PRINT "Ballistics Coefficient  = "; BC
ELSE
GOTO F:
END IF
END


SUB c (koef, m, dm, BC, a)
BC = koef
koef = koef * (1 - dm / m)
END SUB


SUB InfHyres (x0, y0, z0, v0, w0, a, h0, ta, pa, ea, tp, ta1, pa1, xx1, voo,
vo1, wind, koef, cw, vv, E, F, D, Pr, TE, m, dm, atm, G, GA, GS, Tc)
LOCATE 4, 12: INPUT "Atmosphere: ASM = 1, ICAO = 2; Traditional
Standard = 3"; atm
CLS
IF atm = 1 THEN GOTO 100:
IF atm = 2 THEN GOTO 200:
IF atm = 3 THEN GOTO 300:


100 LOCATE 4, 13: INPUT "G-Function; G1= 1, G2=2, G5=5, G6=6,
G7=7, G8=8, Ingalls= 10"; G
TE = 289.6: Pr = 750: Tc = 21.1
IF G = 1 THEN E = .312914: F = 79.3976: D = .000100347#
```

```
IF G = 2 THEN E = .1413: F = 29.9097: D = .00009116233#
IF G = 5 THEN E = .19734: F = 49.0806: D = .0000743571#
IF G = 6 THEN E = .140533: F = 23.6633: D = .00010334#
IF G = 7 THEN E = .150355: F = 34.7319: D = .000056648#
IF G = 8 THEN E = .14733: F = 29.085: D = .000099366#
IF G = 10 THEN E = .32072: F = 82.6909: D = .000108774#
CLS
GOTO 400:

200 LOCATE 4, 13: INPUT "G-Function; G1= 1, G2=2, G5=5, G6=6,
G7=7, G8=8, Ingalls = 10"; GA
TE = 288.15: Pr = 760: Tc = 21.1
IF GA = 1 THEN E = .31574: F = 78.6769: D = .00010584#
IF GA = 2 THEN E = .143353: F = 30.2415: D = .00009287#
IF GA = 5 THEN E = .200207: F = 49.625: D = .000075244#
IF GA = 6 THEN E = .142352: F = 23.6937: D = .000105244#
IF GA = 7 THEN E = .152593: F = 35.1717: D = .000057679#
IF GA = 8 THEN E = .149441: F = 29.379: D = .000101154#
IF GA = 10 THEN E = .325383: F = 83.6082: D = .00010724#
CLS
GOTO 400:

300 LOCATE 4, 13: INPUT "G-Function; Siacci = 1, Mayevski = 2, G43-
Function =3"; GS
TE = 289.08: Pr = 750: Tc = 15
IF GS = 1 THEN E = .3333333#: F = 80: D = .000212
IF GS = 2 THEN E = .320243: F = 81.3721: D = .00010807#
IF GS = 3 THEN E = .157713: F = 36.39542: D = .00007454#
CLS
GOTO 400:

400


LOCATE 4, 13: INPUT "Departure Angle [Degree]          = "; z0
LOCATE 5, 13: INPUT "Departure Speed [m/s]             = "; y0
LOCATE 6, 13: INPUT "Temperature of Air [C] at firing site  = "; ta
LOCATE 7, 13: INPUT "Propellant Temperature[C]           = "; tp
LOCATE 8, 13: INPUT "Pressure [mm] at the firing site     = "; pa
LOCATE 9, 13: INPUT "Pressure of Vapor [mm] at firing site  = "; ea
LOCATE 10, 13: INPUT "Mass of Projectile              = "; m
LOCATE 11, 13: INPUT "Change in Projectile Mass         = "; dm
```

```
LOCATE 12, 13: INPUT "Range Wind                    = "; wind
LOCATE 13, 13: INPUT "Cross Wind                    = "; cw
LOCATE 14, 13: INPUT "Ballistics Coefficient          = "; koef
LOCATE 15, 13: INPUT "x-coordinate of a point on Trajectory = "; xx1
LOCATE 16, 13: INPUT "Integration Step, 10, 1, or 0.5, 0.1 = "; h0
vv = v0: a = z0: voo = y0
ta = ta + 273.15
pal = ta / (1 - .3785 * ea / pa)
vol = (voo - .4 * voo * (dm / m) + .0013 * voo * (tp - Tc))
y0 = SQR(vol ^ 2 + wind ^ 2 - 2 * vol * wind * COS(a * 3.141592654# /
180))
y0 = y0 * COS(a * 3.141592654# / 180)
z0 = TAN(a * 3.141592654# / 180)
z0 = z0 / (1 - wind / (vol * COS(a * 3.141592654# / 180)))
CLS
END SUB

SUB menu (cog, cof, xf, yf, xfu, yfu, t$)
COLOR cog, cof
LOCATE xf - 1, yf: PRINT t$
LOCATE xf, yf: PRINT "É" + STRING$(yfu - yf, 205) + "»";
FOR i = xf + 1 TO xfu
LOCATE i, yf: PRINT "°" + SPACE$(yfu - yf) + "°";
NEXT
LOCATE xfu + 1, yf: PRINT "È" + STRING$(yfu - yf, 205) + "¼";
END SUB

SUB NPkoef (k, L, r, q, h, y1, z1, v1, w1)
k = h * y1: L = h * z1
r = h * v1: q = h * w1
END SUB

SUB NPxyzvw (nk, x, x0, y, y0, z, z0, v, v0, w, w0, h, h0, k, L, r, q)
IF nk = 1 THEN
x = x0: y = y0: z = z0
v = v0: w = w0: h = h0
GOTO fund:
END IF

IF nk = 2 OR nk = 3 THEN
x = x0 + (.5 * h): y = y0 + (.5 * k)
z = z0 + (.5 * L): v = v0 + (.5 * r)
```

```
w = w0 + (.5 * q)
GOTO fund:
END IF
IF nk = 4 THEN
x = x0 + h: y = y0 + k: z = z0 + L
v = v0 + r: w = w0 + q
END IF
fund:
END SUB


SUB y1z1v1w1 (x, y, z, v, w, y1, z1, v1, w1, koef, pa1, wind, ys, yy, pa,
ta1, TE, E, F, D, Pr)
ta1 = (TE / pa1) ^ .5
yy = y * SQR(1 + z ^ 2)
IF yy * ta1 > 256! THEN
y1 = -1 * koef * (pa / Pr) * ((pa1 - .006328 * v) / pa1) ^ 4.4 * (ta1 * E *
yy - F) / yy
ELSE
y1 = -1 * koef * (pa / Pr) * ((pa1 - .006328 * v) / pa1) ^ 4.4 * D * (ta1 *
yy) ^ 2 / yy
END IF
z1 = -9.80665 / y ^ 2
v1 = z
w1 = 1 / y
END SUB
```

2.18. PC Program for Projectile Drop and Inclined Fire

The PC RPOINT.BAS is a Qbasic program that can be used to find the elements of a trajectory at any point, as for example the elements at the impact point for the inclined firing. It can be used as well to find the projectile drop at a given horizontal distance from the launching point when the projectile is fired horizontally in non-standard or standard atmosphere.

The main input data are the departure angle, the departure speed, the x-coordinate of a point on the trajectory, and a fixed ballistics coefficient.

The program can be modified to be used when the ballistics coefficient is a function of the projectile speed or a function of the departure angle.

The following examples and exercises illustrate the use of the program displayed at the end of this section.

Example 18.1 Elements of the Trajectory for a Given x-coordinate

A Russian 122mm projectile is fired from the field cannon Mod 1960 with a speed of 885m/s. Find the elements of the trajectory at the point with x-coordinates 12,200 meters and 14,000 meters if the departure angle is 10 degrees.

The ballistics coefficient as a function of elevation (departure) angle is given in table 1, example 17.9. The table can be considered that represents the ballistics coefficient as a function of the departure angle, i.e., $c = f(\alpha_0)$.

At the firing site, the temperature of dry air and the atmospheric pressure are respectively 8 degrees Celsius, and 750mmHg. The air humidity is 50%.

Assume that there is a ballistics range—wind of 5 m/s and a cross-wind of 3 m/s.

Solution:

Preparatory data

The ballistics coefficient
Using table 1 displayed in example 17.9, by interpolation, we find that the ballistics coefficient is 0.505.

The pressure of water vapors
Using table 1 section 2.2, we find that the pressure of water vapors for the air humidity 50% is 3.935mm Hg.

Input: Atmosphere TSA: 3; G43 function: 3; Initial coordinate: 0; Departure speed: 885; Departure angle: 10; Temperature of air at firing site: 8; Temperature of propellant: 8; Atmospheric pressure at firing site: 750; Pressure of air vapors at firing site: 3.935; Projectile mass 1; Change in projectile mass 0; Range wind: 5; Cross wind: 3. Ballistics coefficient: 0.505; X-coordinate of a point: 12200, 16000; Integration step: 0.5

Output: Range12200m: Corresponding y-coordinate 173.64m; Time of flight23.15s; Terminal speed: 347m/s; Terminal angle: -16.20 degree; Cross wind deflection: 118.73m.

Range 14000m: corresponding y-coordinate: 494.42; time of flight: 28.87s; terminal speed: 318m/s; terminal angle: -25.28 degrees; cross wind deflection: 143.17m.

The coordinates of the trajectory vertex are (7413, 772.34).

Note that the projectile falls on the site that is inclined at an angle:

$$E = \arctan(492.42 / 14000) = 2.01443°$$

Example 18.2 Bullet Drop, G_7 Drag Function

A NATO 5.56mm Ball M855 bullet (62 grain/ 0.004 kg) is fired horizontally with an initial speed of 3100 ft/s (940m/s).

Find the drop of the bullet for the effective ranges of 100-800 meters.

The ballistics coefficient that corresponds to the Army Standard Metro with respect to the G_7-drag function is 0.151lb/in^2.

Find as well the departure angle needed to hit the target at the horizontal range of 300 meters.

At the firing site, the atmosphere is ASM standard. The temperature of the propellant charge is 21.1 degrees Celsius.

Solution:

The ballistics coefficient in metric system of units is

$$c = \frac{1.422}{C} = \frac{1.422}{0.151} = 9.417 m^2 / kg.$$

Using table 1 section 2.2, we find that the air vapor pressure (at 15 degrees Celsius and 78% humidity) is 0.59 mmHg.

Using the RPOINT.BAS, we find the drop of the projectile for the horizontal x-coordinate is 300 meters.

Input: Atmosphere ASM: Press 1; G_7 function: Press 7; Initial coordinate: 0; Departure speed: 940; Departure angle: 0; Temperature of air at firing site: 15; Temperature of propellant: 21.1; Atmospheric pressure at firing site: 750; Pressure of air vapors at firing site: 0.59; Projectile mass 1; Change in projectile mass 0; Range wind:0; Cross wind: 0.
Ballistics coefficient: 9.417; X-coordinate of a point: 300, 500; Integration step: 0.5.

Output: Range 300m: y-coordinate (Drop) is -0.646m; Time of flight is 0.386s; Terminal speed is 641m/s; Terminal angle is -0.2829 degree; Cross wind deflection is 0m

Range 500m: y-coordinate (drop) is -2.22m); time of flight is 0.749; terminal speed is 475m/s; terminal angle is -0.6559 degree; cross wind deflection is 0m.

In the same way, we find the drop of the given projectile for the other horizontal ranges.

Answer:

The projectile drop, for horizontal ranges:

100, 200, 300, 400, 500, 600, 700, 800

is respectively:

-0.06, -0.262, -0.646, -1.272, -2.22, -3.606, -5.585, -8.357.

The departure angle for the horizontal range 300 meters

Using the property of the rigidity of the trajectory (see section 1.5, "*Exterior Ballistics of Small Arms*") we find that the departure angle needed to hit the target at 300 meters is

$$\alpha_0 = \arctan(\frac{drop}{range}) = \arctan(\frac{0.646}{300}) = 0.1234°$$

The PC Program RPOINT.BAS, Projectile Drop, Inclined Fire

```
'FIND : Elements of Trajectory at any two points, as well as the projectile
'          drop when departure angle is zero
'GIVEN: Departure Angle, Initial speed, BC
'-----------------------------------------------------------------------------------
' DATA
' Input: y-coordinate of FIREARM = 0, departure Angle; 0, departure speed
,856.5
' Temperature of Air, 25, temperature of propellant, 25;
'Pressure = 740, Pressure of Air vapor = 6.35, Projectile mass, 1;
'Change in Projectile mass = 0, range wind, 10; cross wind, 0; BC = 3.182
'x-coordinate of a Point= 400; X- coordinate of a Point = 600
'Results: x-coordinate= 400, y-coordinate, -1.3338
'X- coordinate= 600, y-coordinate, - 3.474
'-----------------------------------------------------------------------------------


'Functions & Subs.

DECLARE SUB y1z1v1w1 (x, y, z, v, w, y1, z1, v1, w1, koef, pa1, wind, ys, yy,
pa, ta1, TE, E, F, D, Pr)
DECLARE SUB InfHyres (x0, y0, z0, v0, w0, a, h0, ta, pa, ea, tp, ta1, pa1, xx1,
voo, vo1, wind, koef, cw, vv, E, F, D, Pr, TE, m, dm, xax)
DECLARE SUB NPxyzvw (nk, x, x0, y, y0, z, z0, v, v0, w, w0, h, h0, k, L, r, q)
DECLARE SUB NPkoef (k, L, r, q, h, y1, z1, v1, w1)
DECLARE SUB menu (cog, cof, xf, yf, xfu, yfu, t$)
DECLARE SUB c (koef, m, dm, BC)

'Variables

DIM m(4, 4), v(4)
rendi = 4
cog = 7: cof = 0

'Zgjidhja
CLS

fillimi:
menu cog, cof, 3, 10, 21, 70, "INITIAL DATA"
InfHyres x0, y0, z0, v0, w0, a, h0, ta, pa, ea, tp, ta1, pa1, xx1, voo, vo1, wind,
koef, cw, vv, E, F, D, Pr, TE, m, dm, xax
c koef, m, dm, BC

F:
FOR nk = 1 TO rendi
NPxyzvw nk, x, x0, y, y0, z, z0, v, v0, w, w0, h, h0, k, L, r, q
```

```
y1z1v1w1 x, y, z, v, w, y1, z1, v1, w1, koef, pa1, wind, ys, yy, pa, ta1, TE, E, F,
D, Pr
NPkoef k, L, r, q, h, y1, z1, v1, w1
m(nk, 1) = k: m(nk, 2) = L
m(nk, 3) = r: m(nk, 4) = q
NEXT nk

'Calculation
FOR i = 1 TO rendi
v(i) = 1 / 6 * (m(1, i) + 2 * m(2, i) + 2 * m(3, i) + m(4, i))
NEXT i

'New Data
x0 = x0 + h: y0 = y0 + v(1): z0 = z0 + v(2)
v0 = v0 + v(3): w0 = w0 + v(4)

IF ABS(z0) < .00001 THEN
ymax = v0
xmax = x0 + wind * w0
END IF

xxx = x0 + wind * w0
IF (xxx - xx1) <= .001 THEN
xc = xxx
yc = v0
tc = w0
ac = 180 * ATN(z0) / 3.141592654#
vc = y0 * (1 + z0 ^ 2) ^ .5
zc = cw * (w0 - xc / (voo * COS(a)))
END IF

xxT = x0 + wind * w0
IF (xxT - xax) <= .001 THEN
tt = w0
xt = xxT
yt = v0
at = 180 * ATN(z0) / 3.141592654#
vt = y0 * (1 + z0 ^ 2) ^ .5
zt = cw * (w0 - xt / (voo * COS(a)))
 END IF
ytt = yt
IF x0 >= 10 AND ABS(v0 - ytt) >= .1 THEN
'Display Results
menu cog, cof, 6, 20, 22, 72, "RESULTS:"
LOCATE 7, 25: PRINT "x-coordinate of Point[m]  = "; INT((xc) + .5)
LOCATE 8, 25: PRINT "Coresponding y-Coord [m]  = "; INT((yc) * 1000 + .5) /
1000
```

```
LOCATE 9, 25: PRINT "Departure Angle [Deg.]     = "; a
LOCATE 10, 25: PRINT "Time of Flight [s]      = "; INT((tc) * 1000 + .5) / 1000
LOCATE 11, 25: PRINT "Terminal Speed [m/s]    = "; INT((vc) + .5)
LOCATE 12, 25: PRINT "Terminal Angle [Deg.]   = "; INT((ac) * 10000 + .5) /
10000
LOCATE 13, 25: PRINT "Cross-Wind Deflection   = "; INT((zc) * 100 + .5) /
100
LOCATE 15, 25: PRINT "Second Point[m]         = "; "("; INT((xt) + .5); ",";
INT((yt) * 1000 + .5) / 1000; ")"
LOCATE 16, 25: PRINT "Time [s]                = "; INT((tt) * 1000 + .5) / 1000
LOCATE 17, 25: PRINT "Corresponding Speed [m/s] = "; INT((vt) + .5)
LOCATE 18, 25: PRINT "Corresponding Angle [Deg] = "; INT((at) * 10000 +
.5) / 10000
LOCATE 19, 25: PRINT "Cross-Wind Deflection   = "; INT((zt) * 100 + .5) /
100
LOCATE 20, 25: PRINT "Trajectory Vertex [m]   = "; "("; INT((xmax) + .5);
","; INT((ymax) * 100 + .5) / 100; ")"
LOCATE 22, 25: PRINT "Ballistics Coefficient  = "; BC
ELSE
GOTO F:
END IF
END

SUB c (koef, m, dm, BC)
BC = koef
koef = koef * (1 - dm / m)
END SUB

SUB InfHyres (x0, y0, z0, v0, w0, a, h0, ta, pa, ea, tp, ta1, pa1, xx1, voo, vo1,
wind, koef, cw, vv, E, F, D, Pr, TE, m, dm, xax)
LOCATE 4, 12: INPUT "Atmosphere: ASM = 1, ICAO = 2; Traditional Standard
= 3"; atm
CLS
IF atm = 1 THEN GOTO 100:
IF atm = 2 THEN GOTO 200:
IF atm = 3 THEN GOTO 300:

100 LOCATE 4, 13: INPUT "G-Function; G1= 1, G2=2, G5=5, G6=6, G7=7,
G8=8, Ingalls= 10"; G
TE = 289.6: Pr = 750
IF G = 1 THEN E = .312914: F = 79.3976: D = .000100347#
IF G = 2 THEN E = .1413: F = 29.9097: D = .00009116233#
IF G = 5 THEN E = .19734: F = 49.0806: D = .0000743571#
IF G = 6 THEN E = .140533: F = 23.6633: D = .00010334#
IF G = 7 THEN E = .150355: F = 34.7319: D = .000056648#
IF G = 8 THEN E = .14733: F = 29.085: D = .000099366#
IF G = 10 THEN E = .32072: F = 82.6909: D = .000108774#
```

```
CLS
GOTO 400:

200 LOCATE 4, 13: INPUT "G-Function; G1= 1, G2=2, G5=5, G6=6, G7=7,
G8=8, Ingalls = 10"; GA
TE = 288.15: Pr = 760
IF GA = 1 THEN E = .31574: F = 78.6769: D = .00010584#
IF GA = 2 THEN E = .143353: F = 30.2415: D = .00009287#
IF GA = 5 THEN E = .200207: F = 49.625: D = .000075244#
IF GA = 6 THEN E = .142352: F = 23.6937: D = .000105244#
IF GA = 7 THEN E = .152593: F = 35.1717: D = .000057679#
IF GA = 8 THEN E = .149441: F = 29.379: D = .000101154#
IF GA = 10 THEN E = .325383: F = 83.6082: D = .00010724#
CLS
GOTO 400:
300 LOCATE 4, 13: INPUT "G-Function; Siacci = 1, Mayevski = 2, G43-
Function =3"; GS
TE = 289.08
Pr = 750
IF GS = 1 THEN E = .3333333#: F = 80: D = .000212
IF GS = 2 THEN E = .320243: F = 81.3721: D = .00010807#
IF GS = 3 THEN E = .157713: F = 36.39542: D = .00007454#
CLS
GOTO 400:

400
LOCATE 5, 13: INPUT "y-coordinate of FIREARM      = "; v0
LOCATE 6, 13: INPUT "Departure Speed [m/s]      = "; y0
LOCATE 7, 13: INPUT "Departure Angle [Degree]      = "; z0
LOCATE 8, 13: INPUT "Temperature of Air [C]      = "; ta
LOCATE 9, 13: INPUT "Propellant Temperature[C]     = "; tp
LOCATE 10, 13: INPUT "Atmospheric Pressure [mm]     = "; pa
LOCATE 11, 13: INPUT "Pressure of Air Vapor [mm]    = "; ea
LOCATE 12, 13: INPUT "Projectile Mass        = "; m
LOCATE 13, 13: INPUT "Change in Projectile mass   = "; dm
LOCATE 14, 13: INPUT "Range Wind             = "; wind
LOCATE 15, 13: INPUT "Cross Wind             = "; cw
LOCATE 16, 13: INPUT "x-coordinate of a point on Trajectory = "; xx1
LOCATE 17, 13: INPUT "x-coordinate of a point on Trajectory = "; xax
LOCATE 18, 13: INPUT "Ballistics Coefficient      = "; koef
LOCATE 19, 13: INPUT "Integration Step, 10, 1, or 0.5 = "; h0
vv = v0: a = z0: voo = y0
ta = ta + 273.15
pa1 = ta / (1 - .3785 * ea / pa)
vo1 = (voo - .4 * voo * (dm / m) + .0013 * voo * (tp - 15))
y0 = SQR(vo1 ^ 2 + wind ^ 2 - 2 * vo1 * wind * COS(a * 3.141592654# / 180))
y0 = y0 * COS(a * 3.141592654# / 180)
```

```
z0 = TAN(a * 3.141592654# / 180)
z0 = z0 / (1 - wind / (vo1 * COS(a * 3.141592654# / 180)))
CLS
END SUB

SUB menu (cog, cof, xf, yf, xfu, yfu, t$)
COLOR cog, cof
LOCATE xf - 1, yf: PRINT t$
LOCATE xf, yf: PRINT "É" + STRING$(yfu - yf, 205) + "»";
FOR i = xf + 1 TO xfu
LOCATE i, yf: PRINT "°" + SPACE$(yfu - yf) + "°";
NEXT
LOCATE xfu + 1, yf: PRINT "È" + STRING$(yfu - yf, 205) + "¼";
END SUB

SUB NPkoef (k, L, r, q, h, y1, z1, v1, w1)
k = h * y1: L = h * z1
r = h * v1: q = h * w1
END SUB

SUB NPxyzvw (nk, x, x0, y, y0, z, z0, v, v0, w, w0, h, h0, k, L, r, q)
IF nk = 1 THEN
x = x0: y = y0: z = z0
v = v0: w = w0: h = h0
GOTO fund:
END IF
IF nk = 2 OR nk = 3 THEN
x = x0 + (.5 * h): y = y0 + (.5 * k)
z = z0 + (.5 * L): v = v0 + (.5 * r)
w = w0 + (.5 * q)
GOTO fund:
END IF

IF nk = 4 THEN
x = x0 + h: y = y0 + k: z = z0 + L
v = v0 + r: w = w0 + q
END IF
fund:
END SUB

SUB y1z1v1w1 (x, y, z, v, w, y1, z1, v1, w1, koef, pa1, wind, ys, yy, pa, ta1, TE,
E, F, D, Pr)
ta1 = (TE / pa1) ^ .5
yy = y * SQR(1 + z ^ 2)
IF yy * ta1 > 256! THEN
y1 = -1 * koef * (pa / Pr) * ((pa1 - .006328 * v) / pa1) ^ 4.4 * (ta1 * E * yy - F) /
yy
```

```
ELSE
y1 = -1 * koef * (pa / Pr) * ((pa1 - .006328 * v) / pa1) ^ 4.4 * D * (ta1 * yy) ^ 2 /
yy
END IF
z1 = -9.80665 / y ^ 2
v1 = z
w1 = 1 / y
END SUB
```

2.19. PC Program to Compute the Departure Angle

From the practical point of view, the most important task of
shooting is the determination of the departure angle of a given
projectile in order to hit a given target that is located within the
horizontal or inclined range that can be reached by the firearm.

The computation of the departure angle of a projectile is one of
the main problems in exterior ballistics.

The PC programs APROJC.BAS can be used to find the departure
angle of a projectile fired with a given initial speed when the
coordinates of the firearm and the target, are known.

The ballistics coefficient that is one of the data we input to
execute the PC program, is constant. The APROJC.BAS can be
easily modified to be used in general case when the ballistics
coefficient is a function of the departure angle, or a function of
departure speed of the projectile.

The APROJC.BAS calculates as well the coordinates of the
trajectory vertex, the time of flight to the target, the terminal speed
and the terminal angle and the crosswind deflection of the
projectile.

Using APROJC.BAS, we are able to find the trajectory elements
of a point on the trajectory for a given x-coordinate of that point.
The value of x-coordinate of the point must be less than the
projectile range.

The PC program APROJC.BAS can be used to find the departure
angle for anti-aircraft or uphill shooting, as well as for downhill
shooting.

For up-hill shooting, the coordinates of the trajectory vertex
calculated using APROJC.BAS are (0, 0). That means that the
trajectory vertex "does not exists," and the largest altitude of the
trajectory is at the point of impact. The given PC program is unable
to find the vertex since the determination of the vertex is
programmed to be found as the point on the trajectory where the
tangent is parallel to the x-axis.

But, the program can be modified to find the maximum ordinate with respect to the inclined site.

Example 19.1 Cannon Projectile, G_{43}-Function

The departure speed of a 122mm projectile launched by a Russian field cannon Model 1960 is 885m/s. We assume that the projectile flight is in the TSA atmosphere.

(a) Find the departure angle needed to hit a target located at the sea level at the horizontal range of 10,000m. Assume the absence of wind.
(b) Find the departure angle needed to hit the target located at the sea level at the horizontal range of 10,000m. Consider the flight of projectile in TSA atmosphere at sea level, but in presence of a range wind of 10m/s.

The ballistics coefficient related with the G_{43}-function of resistance is given in table 1, example 17.8.

Solution:

Since we do not know the departure angle, we can use a ballistics coefficient of 0.51 (see table 1, example 17.1) for approximate estimation of the departure angle.

Using APROJC.BAS

(a) **Input**: x-coordinate of target, 10000; y-coordinate of target, 0; y-coordinate of cannon, 0; departure speed, 885m/s; temperature of air, 15; temperature of propellant, 15; atmospheric pressure, 750; pressure of air vapors, 6.35; projectile mass, 1; change in projectile mass, 0; range wind, 0; cross wind, 0; BC, 0.51.

Output: Departure angle 6.32813 Degree.

Using the approximate value of 6.32813 degrees of the departure angle and the table 1 of example 17.1, we find a new ballistics coefficient 0.498.

Employing again the PC program APROJC.BAS, but with the BC = 0.498, we find that the departure angle is 6.22607 degree.

Using table 1 of example 17.8, we find an average ballistics coefficient of the ballistics coefficient as a function of the departure angle is given in.

(b) In the same way as in (a), but inputting the range wind 10m/s, and the ballistics coefficient 0.498, we find that the departure angle is 6.14417 degree.

Example Bullet Departure Angle, G_7-Function

A NATO 5.56mm Ball M855 bullet (62 grain/ 0.004 kg) is fired horizontally with an initial speed 940m/s.

Compute the departure angle needed to hit the target at the horizontal range of 300 meters.

The ballistics coefficient that corresponds to the Army Standard Metro with respect to the G_7-drag function is 9.147m^2/kg (see example 18.2).

Solution:

Input: x-coordinate of target, 300; y-coordinate of target, 0; y-coordinate of cannon, 0; departure speed, 940m/s; temperature of air, 15; temperature of propellant, 15; atmospheric pressure, 750; pressure of air vapors, 6.35; projectile mass, 1; change in projectile mass, 0; range wind, 0; cross wind, 0; BC, 9.147.

Output: departure angle 0.12250 degree; Time of flight 0.384s; Terminal speed: 648m/s; Terminal angle: -0.1558 degree; Cross wind deflection: 0m; vertex (163, 0.18)

PC Program APROJC.BAS, Departure Angle

```
'Ballistics Coefficient is known
'Non Standard Atmosphere, Wind Present
'Find:  Departure Angle, Time of Flight, Impact Speed, Impact Angle, etc.
'Given: The coordinates of the target and the location of the muzzle of the
cannon
'  & Ballistics Coefficient
' Temperature of air and thrusting charge of projectile are known;
'  The weight of projectile and the air humidity are known
' Projectile flight is in presence of wind.
'-------------------------------------------------------------------------------
' CONTROL DATA
' INPUT
' x0 = 10,000,  projectile speed = 885
' Temperature of air = 15,Temperature of propellant= 15
' Pressure = 750, Pressure of Air vapor = 6.35, projectile mass = 27.3,
' change in projectile mass = 0, range wind = 10, cross wind, 0.
'
' RESULTS
' Departure Angle = 6.12 Degree, Time of Flight = 16.6s
' Impact Speed = 419m/s, Impact Angle = -10.106 Degree
' Coordinates of the trajectory vertex (5610, 339)
'-------------------------------------------------------------------------------
'Functions and Sub. Prog.
DECLARE SUB y1z1v1w1 (x, y, z, v, w, y1, z1, v1, w1, koef, pa1, wind, ys, yy,
pa, ta1, TE, Pr, E, F, D)
DECLARE SUB InfHyres (xx, voo, vo1, ta, ta1, pa, pa1, ea, m, dm, tp, wind, xc,
yc, vo, yc1, cw, xx1, koef, atm, G, GA, GS, TE, Pr, E, F, D)
DECLARE SUB InfDales (x0, y0, z0, v0, w0, xc, yc, A, aT, vo, yc1, xm, ym, cw,
xx, voo, xa, ya, ta, va, aa)
DECLARE SUB NPxyzvw (nk, x, x0, y, y0, z, z0, v, v0, w, w0, h, h0, k, l, r, q)
DECLARE SUB NPkoef (k, l, r, q, h, y1, z1, v1, w1)
DECLARE SUB menu (cog, cof, xf, yf, xfu, yfu, t$)
DECLARE SUB y0z0 (y0, z0, A, vo1, wind)
DECLARE SUB c (koef)
'Variables
SCREEN 0
1:
DIM m(4, 4), v(4)      'Intermediate values (k,l,r,q)
rendi = 4              'rend dif.
cog = 7: cof = 0
cikli = 0
A = 23                 'Initial Angle 23 degree
kendi = 22              'Angle [Degree] for maximum distance
kov = 1                'Test of the value of v0
```

```
gab = 1                    'error 0.1 m.
tt = 1

'Solution
CLS

'Initial Data
menu cog, cof, 3, 10, 7, 70, "DATA INPUT"
InfHyres xx, voo, vol, ta, ta1, pa, pa1, ea, m, dm, tp, wind, xc, yc, yc1, vo, cw,
xx1, koef, atm, G, GA, GS, TE, Pr, E, F, D
hap = 1
'Initial values

F:
x0 = 0: v0 = vo: w0 = 0
y0z0 y0, z0, A, vol, wind: h0 = hap
c koef

ff:
FOR nk = 1 TO rendi
NPxyzvw nk, x, x0, y, y0, z, z0, v, v0, w, w0, h, h0, k, l, r, q
y1z1v1w1 x, y, z, v, w, y1, z1, v1, w1, koef, pa1, wind, ys, yy, pa, ta1, TE, Pr, E,
F, D
NPkoef k, l, r, q, h, y1, z1, v1, w1
m(nk, 1) = k: m(nk, 2) = l
m(nk, 3) = r: m(nk, 4) = q
NEXT nk

'Estimation for new points
FOR i = 1 TO rendi
v(i) = 1 / 6 * (m(1, i) + 2 * m(2, i) + 2 * m(3, i) + m(4, i))
NEXT i

'New Points
x0 = x0 + h: y0 = y0 + v(1): z0 = z0 + v(2)
v0 = v0 + v(3): w0 = w0 + v(4)
xmm = x0 + wind * w0
IF ABS(z0) <= .0001 THEN
xm = xmm
ym = v0
END IF

xaa = x0 + wind * w0
IF ABS(xx1 - xaa) <= 1 THEN
xa = xaa
ya = v0
ta = w0
```

```
va = y0 / COS(ATN(z0))
aa = (180 / 3.1415169542#) * ATN(z0)
END IF


'Tests the y-value
xT = x0 + wind * w0
IF kov = 1 THEN kov = -1: GOTO ff:
IF ABS(xT - xc) < gab AND ABS(v0 - yc) <= (gab * TAN(A * 3.1415954# /
180)) THEN
c:
'DISPLAY of RESULTS
CLS
PLAY "a8a16a32b8"
menu 12, 0, 5, 10, 11, 70, "RESULTS:"
COLOR 7
InfDales x0, y0, z0, v0, w0, A, xc, yc, aT, vo, yc1, xm, ym, cw, xx, voo, xa, ya,
ta, va, aa
CLS
GOTO 1:
END IF


IF ABS(xT - xc) < gab AND (v0 - yc) > (gab * TAN(A * 3.1415954# / 180))
THEN
t$ = "  * ? *"
menu 18, 0, 10, 20, 14, 60, t$
COLOR 14
LOCATE 12, 30: PRINT "Wait a moment, Please (+)";
LOCATE 12, 53: PRINT tt
tt = tt + 1
COLOR 7
A = A - kendi
GOTO fff:
END IF


IF ABS(xT - xc) < gab AND (v0 - yc) < ((-1 * gab) * TAN(A * 3.1415954# /
180)) THEN
t$ = "   * ? *"
menu 18, 0, 10, 20, 14, 60, t$
COLOR 14
LOCATE 12, 30: PRINT "Wait a moment, Please (-)";
LOCATE 12, 53: PRINT tt
tt = tt + 1
COLOR 7
A = A + kendi
GOTO fff:
END IF
GOTO ff:
```

```
fff:
'Restart Cycle
cikli = cikli + 1
IF cikli = 20 THEN GOTO c:
kendi = kendi / 2
kov = 1
GOTO F:

SUB c (koef)
koef = koef
END SUB

SUB InfDales (x0, y0, z0, v0, w0, A, xc, aT, yc, vo, yc1, xm, ym, cw, xx, voo,
xa, ya, ta, va, aa)

aT = ATN(z0) * 180 / 3.141592654#
LOCATE 6, 16: PRINT "Departure Angle          :"; A; "Degree"
LOCATE 7, 16: PRINT "Time of Flight to Target   :"; INT((w0) * 1000 + .5) /
1000
LOCATE 8, 16: PRINT "Terminal Speed             :"; INT((y0 / COS(ATN(z0))) *
100 + .5) / 100; " m/s"
LOCATE 9, 16: PRINT "Terminal Angle           :"; aT; " degree"
LOCATE 10, 16: PRINT "Coordinates of Vertex    :"; "("; INT((xm) * 100 + .5)
/ 100; ","; INT((ym) * 100 + .5) / 100; ")"
LOCATE 11, 16: PRINT "Cross-Wind Deflection      :"; INT((cw * (w0 - xx /
(voo * COS(A * 3.14159265# / 180)))) * 1000 + .5) / 1000
LOCATE 13, 16: PRINT "Location of TARGET      :"; "("; xx; yc1; ")"
LOCATE 14, 16: PRINT "Location of FIREARM     :"; "(0,"; vo; ")"
LOCATE 15, 16: PRINT "Distance to TARGET      :"; (xx ^ 2 + (yc1 - vo) ^ 2)
^ .5
LOCATE 17, 16: PRINT "X-Coordinate of the Point   :"; INT((xa) * 100 + .5) /
100
LOCATE 18, 16: PRINT "Y-Coordinate of the Point   :"; INT((ya) * 100 + .5) /
100
LOCATE 19, 16: PRINT "Time of Flight to the Point :"; INT((ta) * 100 + .5) /
100
LOCATE 20, 16: PRINT "Speed at the Point        :"; INT((va) * 100 + .5) / 100
LOCATE 21, 16: PRINT "Angle at the Point        :"; aa
LOCATE 22, 16: PRINT "Cross Wind Deflection       :"; INT((cw * (ta - xa /
(voo * COS(A * 3.14159265# / 180)))) * 1000 + .5) / 1000

COLOR 7
LOCATE 24, 11: PRINT " Pres [ P ] to repeat [ Esc ] to end ";
cc$ = INPUT$(1)
IF cc$ = CHR$(27) THEN SCREEN 9: CLS : END
END SUB
```

```
SUB InfHyres (xx, voo, vo1, ta, ta1, pa, pa1, ea, m, dm, tp, wind, xc, yc, yc1, vo,
cw, xx1, koef, atm, G, GA, GS, TE, Pr, E, F, D)
LOCATE 4, 12: INPUT "Atmosphere: ASM = 1, ICAO = 2; Traditional Standard
= 3"; atm
CLS
IF atm = 1 THEN GOTO 100:
IF atm = 2 THEN GOTO 200:
IF atm = 3 THEN GOTO 300:

100 LOCATE 4, 13: INPUT "G-Function; G1= 1, G2=2, G5=5, G6=6, G7=7,
G8=8, Ingalls= 10"; G
TE = 289.6
Pr = 750
IF G = 1 THEN E = .312914: F = 79.3976: D = .000100347#
IF G = 2 THEN E = .1413: F = 29.9097: D = .00009116233#
IF G = 5 THEN E = .19734: F = 49.0806: D = .0000743571#
IF G = 6 THEN E = .140533: F = 23.6633: D = .00010334#
IF G = 7 THEN E = .150355: F = 34.7319: D = .000056648#
IF G = 8 THEN E = .14733: F = 29.085: D = .000099366#
IF G = 10 THEN E = .32072: F = 82.6909: D = .000108774#
CLS
GOTO 400:

200 LOCATE 4, 13: INPUT "G-Function; G1= 1, G2=2, G5=5, G6=6, G7=7,
G8=8, Ingalls = 10"; GA
TE = 288.15
Pr = 760
IF GA = 1 THEN E = .31574: F = 78.6769: D = .00010584#
IF GA = 2 THEN E = .143353: F = 30.2415: D = .00009287#
IF GA = 5 THEN E = .200207: F = 49.625: D = .000075244#
IF GA = 6 THEN E = .142352: F = 23.6937: D = .000105244#
IF GA = 7 THEN E = .152593: F = 35.1717: D = .000057679#
IF GA = 8 THEN E = .149441: F = 29.379: D = .000101154#
IF GA = 10 THEN E = .325383: F = 83.6082: D = .00010724#
CLS
GOTO 400:

300 LOCATE 4, 13: INPUT "G-Function; Siacci = 1, Mayevski = 2, G43-
Function =3"; GS
TE = 289.08
Pr = 750
IF GS = 1 THEN E = .3333333#: F = 80: D = .000212
IF GS = 2 THEN E = .320243: F = 81.3721: D = .00010807#
IF GS = 3 THEN E = .157713: F = 36.39542: D = .00007454#
CLS
GOTO 400:
```

```
400
LOCATE 4, 13: INPUT "x-coordinate of TARGET [m]     = "; xx
        xc = xx
LOCATE 5, 13: INPUT "y-Coordinate of TARGET [m]     = "; yc
        yc1 = yc
LOCATE 6, 13: INPUT "y-Coordinate of FIREARM [m]   = "; vo
LOCATE 7, 13: INPUT "Departure Speed [m/s]          = "; voo
LOCATE 9, 13: INPUT "Temperature of Air [C]        = "; ta
LOCATE 10, 13: INPUT "Propellant Temperature [C]    = "; tp
LOCATE 11, 13: INPUT "Atmospheric Pressure [mm Hg]  = "; pa
LOCATE 12, 13: INPUT "Pressure of Air Vapors [mm Hg]= "; ea
LOCATE 13, 13: INPUT "Projectile Standard Mass [kg] = "; m
LOCATE 14, 13: INPUT "Change in Projectile Mass     = "; dm
LOCATE 15, 13: INPUT "Range-Wind [m/s]             = "; wind
LOCATE 16, 13: INPUT "Cross-Wind [m/s]            = "; cw
LOCATE 17, 13: INPUT "X-Coordinate of a Point [m]   = "; xx1
LOCATE 18, 13: INPUT "Ballistics Coefficient BC    = "; koef
CLS

ta = ta + 273.15
pal = ta / (1 - .3785 * ea / pa)
vo1 = (voo - .4 * voo * (dm / m) + .001 * voo * (tp - 15))
koef = koef * (1 - dm / m)
END SUB

SUB menu (cog, cof, xf, yf, xfu, yfu, t$)
COLOR cog, cof
LOCATE xf - 1, yf: PRINT t$
LOCATE xf, yf: PRINT "É" + STRING$(yfu - yf, 205) + "»";
FOR i = xf + 1 TO xfu
LOCATE i, yf: PRINT "º" + SPACE$(yfu - yf) + "º";
NEXT
LOCATE xfu + 1, yf: PRINT "È" + STRING$(yfu - yf, 205) + "¼";
END SUB

SUB NPkoef (k, l, r, q, h, y1, z1, v1, w1)
k = h * y1: l = h * z1
r = h * v1: q = h * w1
END SUB

SUB NPxyzvw (nk, x, x0, y, y0, z, z0, v, v0, w, w0, h, h0, k, l, r, q)
IF nk = 1 THEN
x = x0: y = y0: z = z0
v = v0: w = w0: h = h0
GOTO fund:
END IF
```

```
IF nk = 2 OR nk = 3 THEN
x = x0 + (.5 * h): y = y0 + (.5 * k)
z = z0 + (.5 * l): v = v0 + (.5 * r)
w = w0 + (.5 * q)
GOTO fund:
END IF

IF nk = 4 THEN
x = x0 + h: y = y0 + k: z = z0 + l
v = v0 + r: w = w0 + q
END IF
fund:
END SUB

SUB y0z0 (y0, z0, A, vo1, wind)
y0 = SQR(vo1 ^ 2 + wind ^ 2 - 2 * vo1 * wind * COS(A * 3.141592654# / 180))
y0 = y0 * COS(A * 3.141592654# / 180)
z0 = TAN(A * 3.141592654# / 180)
z0 = z0 / (1 - wind / (vo1 * COS(A * 3.141592654# / 180)))
END SUB

SUB y1z1v1w1 (x, y, z, v, w, y1, z1, v1, w1, koef, pa1, wind, ys, yy, pa, ta1, TE,
Pr, E, F, D)
ta1 = (TE / pa1) ^ .5
yy = y * SQR(1 + z ^ 2)
IF yy * ta1 > 256! THEN
y1 = -1 * koef * (pa / Pr) * ((pa1 - .006328 * v) / pa1) ^ 4.4 * (ta1 * E * yy - F) /
yy
ELSE
y1 = -1 * koef * (pa / Pr) * ((pa1 - .006328 * v) / pa1) ^ 4.4 * D * (ta1 * yy) ^ 2 /
yy
END IF
z1 = -9.80665 / y ^ 2
v1 = z
w1 = 1 / y
END SUB
```

2.20. Construction of Ballistics Tables

Ballistics tables display the projectile range as a function of initial speed, initial angle, and ballistics coefficient of projectile in tabular form.

The ballistics tables display the horizontal range x_T in standard atmosphere as a function of projectile departure speed v_0, departure angle α_0, and ballistics coefficient c.

The ballistics tables are obtained solving the differential equations of projectile flight, for example system (2.1.4), in standard atmosphere for some sets of values (v_0, α_0, c).

Using ballistics tables and interpolation procedures, we are able to find the trajectory of flight of any given projectile that have particular ballistics characteristics (v_0, α_0, c), and so to construct the range tables.

The ballistics tables are used as well to find the ballistics' main corrections coefficients (see sections 3.4 and 3.5).

The ballistics tables are result of a tremendous and time-consuming work of calculations and time consuming done by the ballisticians of the twentieth20[th] century.

Nowadays, the ballistics tables can be considered obsolete, since the use of high-speed computers assist the ballisticians to integrate numerically in real time the systems of differential equations that describes the projectile flight.

Anyway, we are showing the way we can obtain the ballistics tables that have played a determinant role in the construction of standard range tables of different firearms.

The example 20.1 demonstrates the use of the differential equations of the projectile flight to construct the ballistics tables, while the example 20.2 illustrates the use of the ballistics tables to construct range tables.

Example 20.1 Construction of Ballistics Tables

Find the elements of a ballistics table corresponding to the following sets of ballistics data:

$$c = 0.55, \ v_0 = 850m/s, \ \alpha_0 = 5°$$
$$c = 0.55, \ v_0 = 900m/s, \ \alpha_0 = 5°$$
$$c = 0.60, \ v_0 = 850m/s, \ \alpha_0 = 5°$$
$$c = 0.60, \ v_0 = 900m/s, \ \alpha_0 = 5°$$
$$c = 0.55, \ v_0 = 850m/s, \ \alpha_0 = 10°$$
$$c = 0.60, \ v_0 = 850m/s, \ \alpha_0 = 10°$$
$$c = 0.55, \ v_0 = 900m/s, \ \alpha_0 = 10°$$
$$c = 0.60, \ v_0 = 900m/s, \ \alpha_0 = 10°$$

Solution:

The data presented in the following table are obtained using RangeC.Bas only for the first set of initial values. The corresponding trajectory elements of the other 7seven sets can be obtained in the same way.

Input: $c = 0.55, \ v_0 = 850m/s, \ \alpha_0 = 5°$

Output: Range, 5795m; time of flight, 11.84s; terminal speed, 314.19m/s; terminal angle, -9.69159 degree; Trajectory vertex (3393m, 177.44m).

The partial ballistics table that follows is constructed only for the range of fire that corresponds to the above 8eight sets of initial values.

Table 1: Elements of a Ballistics Table (Siacci G-function, TSA atmosphere)

c_0 \ v_0	$\alpha_0 = 5°$		$\alpha_0 = 10°$	
	850	900	850	900
0.55	5795	6199	8086	8545
0.60	5551	5926	7712	8136

Example 20.2 Use of Ballistics Tables to Construct Range Tables

Use the ballistics table obtained in example 2 to find the range of a projectile fired with an initial speed of 870m/s, and departure angle of 6.4 degrees, if the ballistics coefficient of the projectile is 0.58.

Solution:

The procedure of interpolation of the data of table 1 is presented in the following table.

v_0 / c_0	$\alpha_0 = 5°$			$\alpha_0 = 6.40°$	$\alpha_0 = 10°$		
	850	870	900	870	850	870	900
0.55	5795		6199		8086		8545
0.58	5630.6	5792.44	6035.2	**6431.04**	7922.2	8073.16	8299.6
0.60	5551		5926		7712		8136

Using the interpolation, we have obtained a range of 6431.04m.

A more accurate value, 6537m, for the horizontal range, is obtained using the PC program RangeC.Bas.

The range, estimated using the ballistics table, has a relatively large error of (6537-6431) = 106m.

2.21. Construction of Standard Range Tables

Standard firing range tables are constructed using the data obtained by firing tests that traditionally for the artillery fire are performed for angles 5, 15, 25, 35, and 45 degrees (the largest angle depends on the maximum departure angle).

For small arms, the ballistics coefficient function is determined for horizontal ranges: 500m, 800m, 1,500m, 2,000m, 3,000m, and 4,000m.

In practice, for a given angle (let's say 15 degrees), to find the ballistics coefficient function, the ballisticians fire a certain number of projectiles. The experimental data are being manipulated statistically to determine horizontal range, departure speed, and time of flight.

Normally the other elements of the trajectory are found, theoretically solving the differential equations of projectile flight.

To construct the firing range tables for a given projectile, we determine first the ballistics coefficient as a function of the departure angle, $c = f(\alpha_0)$. Once the function $c = f(\alpha_0)$ is determined, by interpolation, we find the ballistics coefficient for any departure angle.

Then using the PC program APROJC.BAS, we find the elements of the trajectory for any horizontal range in standard conditions.

To find the ballistics coefficient function, $c = f(\alpha_0)$, based on the experimental data, we use the PC program BCPROJ.BAS.

The BC calculated employing BCPROJ.BAS is the BC that corresponds to the given departure angle no matter if the atmospheric conditions are standard or not.

In practice, the parameters that deviate from the standard values (because we do not have control on) are the atmospheric conditions and wind, while the ballistics characteristics of the projectile and firearm can be controlled to have standard values, or close.

The following example demonstrates the use of BCPROJ.BAS and APROJC.BAS to construct the standard range tables.

The method we use does not require converting the experimental data into the standard atmosphere like traditional methods (Siacci's method, etc.) usually require.

Example 21.1 Range Tables Constructed Using PC Programs

Use the G_1-function related with the ICAO atmosphere to construct the range tables of the 122mm Russian projectile (projectile mass is 27.3kg) fired with a speed of 885 m/s.

The projectile flight is in TSA atmosphere: the temperature is 15 degrees Celsius; pressure is 750mm Hg; pressure of water vapors is 6.35mm Hg.

The experimental results are performed in the following conditions:

At the time of firing tests, the values of characteristics of atmosphere were: Temperature 20 degrees Celsius (temperature), pressure 760mm Hg (pressure), pressure of water vapors and 8mm Hg (pressure of water vapors).

Temperature of the propellant charge (15 degrees Celsius), as well as the other ballistics characteristics of the projectile and the propellant, was kept standard.

Range wind was not present (the shooting was performed in the direction perpendicular to the ballistics wind).

The results of the firing tests are presented in the following table.

Table 1:122mm projectile, departure speed 885m/s.

Angle, α_0	5°	15°	25°	35°	45°
x_T	8725	15610	19660	22370	23750

Use the PC programs BCPROJ.BAS and APROJC.BAS

(a) Determine the ballistics coefficient function, $c = f(\alpha_0)$ related with the G_1 -function in ICAO atmosphere and present that BC function in a tabular form;.

(b) Use the $\dot{G}_1$-function of resistance (related with ICAO atmosphere) to calculate the elements of the projectile trajectory for the horizontal range of 12,200 meters in ICAO atmosphere.

Solution:

(a) Using BCPROJ.BAS, Departure angle is 15 degrees, and horizontal range is 15,610m:

Input: ICAO Atmosphere?, press 2, G_1-function?, press 1; Y/N? Press N and Enter, or just Enter; Departure angle? Press 15; Y-Target? Press Enter; y-Gun? Press Enter; Error in x? Press 0.5; Temperature of air? Press 20; Temperature of Propellant? 15; Pressure? Press 760; Pressure of water vapors? Press 8; Projectile mass? Press 27.3; Projectile change in masss? Press 0; range wind? Press 0.

Output: BC = 0.2795

In a similar way, using BCPROJ.BAS, we find the ballistics coefficient that corresponds to the other angles.
The results of calculation are displayed in table 2.

Table 2: Approximate Ballistics Coefficient Function

Angle, α_0	5°	15°	25°	35°	45°
$c=f(\alpha_0)$	0.263	0.2795	0.2846	0.2903	0.2971

(a) To find the elements of the trajectory in standard atmosphere and construct the standard firing tables, we use the PC program APROJC.BAS and the function $c=f(\alpha_0)$ presented in table 2.

We need to determine the departure angles that correspond to the set of horizontal ranges:

200, 400, 600, ... 1,000, 1,200, 1,400 ... 10,000, 10200, ... 23000, 23200 ... 23,800.

We will demonstrate the use of APROJC.Bas for the horizontal range of 12200m.

As a first approach, for the range of 12200m, we use the ballistics coefficient that corresponds to the angle of departure 15 degree, i.e., BC = 0.2795.

Executing APROJC.Bas for the horizontal range of 12,200m, and BC 0.2795, we find the departure angle at 9.26343 degrees.

Using table 2, by interpolation, we find that the corresponding ballistics coefficient is 0.2700.

Executing again APROJC.BAS (range of 12,200, BC 0.2700) we find that the departure angle is at 8.98682 degrees. Using table 2, by interpolation we find BC = 0.2696.

Using again the PC program APROJC.BAS, (range of 12200, BC = 0.2696) we obtain the elements of the trajectory of flight:

Range 12200, BC = 0.2696:

Departure angle is 8.9761 degrees; time of flight is 22.82; terminal speed is 361m/s; terminal angle is -16.2331 degrees; trajectory vertex coordinates are (7041m, 652m).

In the following table, for comparison, are displayed the data obtained using PC programs: Run122.Bas;, Aprojc.Bas (G_{43}-function, BC = 0.5031 derived using table 1, example 17.9), and the PC program of Jan Krecmar's, Ballistica22 (G_{43} -function, and G_1-function).

In table 4, there are included as well the data obtained using the range table of 122mm projectile fired with initial speed 885 m/s from the 122mm Russian cannon, model 1960.

Table 4: Range Table: Range 12,200 meters, ICAO atmosphere

Program	BC	Angle	Range	Time	Speed	Height
Example21.1, G_1	0.2696	8.9761	12200	22.82	361	652
Range Table, G_{43}	0.5031	8.8833	12200	23.00	350	649
ACA122.Bas, Siacci	0.2427	8.8767	12200	22.7	361	643
APROJC.BAS; G_{43}	0.5031	8.9063	12200	22.80	356	647
Ballistica22, G_{43}	0.5031	9.0155	12200	23	345	N/A
Ballistica22, G_1	0.2696	8.9599	12200	22.84	359	N/A

2.22. General Projectile Trajectory-Streamline and Snell's Law Model

The projectile trajectory-streamline and Snell's law model is valid as well for two trajectories that have different ballistics coefficients, for example, as result of changes that are related directly with the ballistics coefficient.

Indeed, consider the differential equation (2.4.6), i.e.,

$$\frac{dv}{dt} = \frac{\rho_0 \cdot a_{0N}}{\rho_{0N} \cdot a_0} \cdot c \cdot h_0 \cdot G_D(v) . \qquad (2.22.7)$$

If we formally replace in it the parameter $(\rho_0 a_{0N} / \rho_{0N} a_0)$ with a "similarity parameter" J defined by the relation

$$X_T = J \cdot x_T , \qquad (2.22.8)$$

we can write

$$\frac{dv}{dt} = J \cdot h_0 \cdot G_D(v) . \qquad (2.22.9)$$

Looking at the equation (2.22.9), we are not able to tell the history of the similarity parameter J, or what the similarity parameter J represents.

Since the differential equation (2.4.6) is the basic equations for the projectile-streamline method developed in sections (2.4) and (2.7), it is obvious that the method can be applied in the general case when both the atmospheric characteristics and the ballistics coefficients of two similitude trajectories are different.

Note that the general projectile trajectory-streamline and Snell's law model can be applied for two projectiles that have different ballistics coefficient, as well as for the same projectile when there is a change in the mass of a given projectile.

For example, a small change dm in the mass of a standard projectile causes a change dc in the corresponding ballistics coefficient ($c = 1000 \cdot id^2 / m$) that in average is

$$dc = -c \cdot \frac{dm}{m}.$$
(2.22.10)

The ballistics coefficient becomes

$$\bar{c} = c \cdot (1 - \frac{dm}{m}),$$
(2.22.11)

while the parameter J is

$$J = \frac{\bar{c}}{c} = 1 - \frac{dm}{m}.$$
(2.22.12)

In the following section, we apply the general projectile trajectory-streamline and Snell's law for the inclined fire.

The illustration examples shows that the general projectile trajectory-streamline and Snell's law model gives acceptable results.

2.23. Construction of Inclined Range Tables

In general, the standard range tables can not be used for the inclined fire since for relatively large angles of sight, the errors are significant.

The approximate methods to fire against a target located on an inclined site are based on the principle of rigidity of the trajectory, [49] which states that the form of the trajectory does not change if it moves up or down at an angle of site A_S that satisfies the relation (figure 5, figure 6):

$$A_S \leq \arctan(20 / x_T),$$
(2.23.1)

where x_T is the horizontal range.

---

49.   Field Artillery , Vol. 6, "Ballistics and Ammunition", 1992, DND Canada

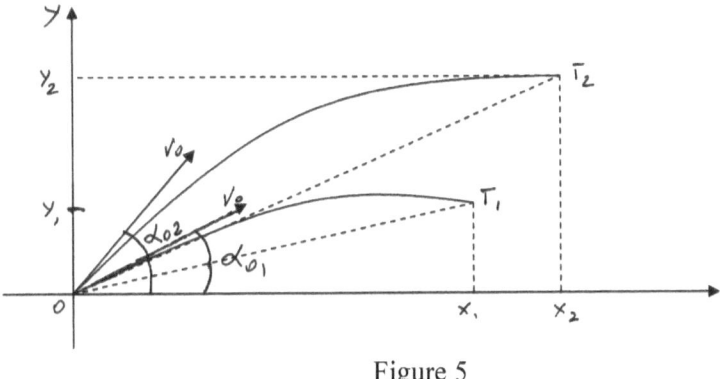

Figure 5

Thus, for such inclined angle of sites, we can use the angle of departure, set up for a given horizontal range, to fire against a target that is on the inclined site at the same range.

For relatively large angles of sighte that do not satisfy (2.23.1), the use of the rigidity principle based on the horizontal range tables gives relatively large errors (see example 23.1).

Firing against a target located on the inclined plane in standard atmosphere is done using:

- the complementary angle of sight corrections;
- the complementary range corrections.

The corrections can be found using the methods presented briefly in Shapiro's *Exterior Ballistics* book. [50].

Hereafter, we show a simple method based on the projectile similitude and Snell's law model.
We assume that the standard horizontal range for a given firearm and projectile is already known, constructed using the ballistics coefficient function

$$c = f(\alpha_0), \qquad\qquad (2.23.2)$$

---

50. Shapiro, J. M. "Exterior Ballistics", p. 286-287, Oborongiz, 50'

that is determined experimentally.

The target is located on the inclined plane that forms an angle of site A_S with the horizon (figure 6).

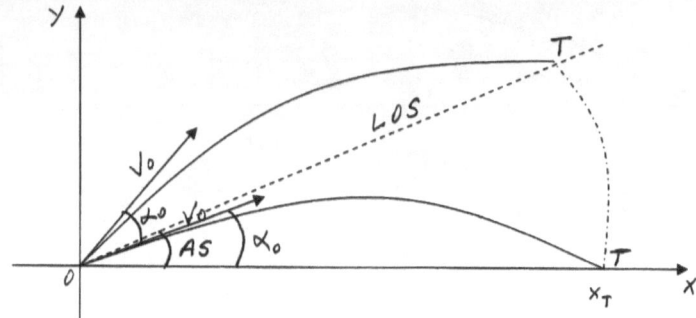

We will use the projectile-streamline and Snell's law model to find the departure angle needed to hit a target on the inclined plane.

For the inclined fire, for two similitude trajectories, we assume that the density, virtual temperature, and pressure are the same. [51]

We denote C_T the ballistics coefficient that corresponds to the trajectory of flight that passes through the point ($x_T,\ y_T$) on the inclined plane where the target is located and c_T the ballistics coefficient of the trajectory that corresponds to the horizontal range x_H.

As a first approach, using the principle of rigidity, we assume that the angle of projection A_p is equal to the departure angle α_0 that corresponds to the horizontal range x_H. We assume as well that the ballistics coefficient of the rigid trajectory is the same as the ballistics coefficient that corresponds to the horizontal range, i.e., to

$$c_T = f(\alpha_0). \qquad (2.23.3)$$

---

51. Field Artillery, Vol. 6, "Ballistics and Ammunition", p.338 , 1992, DND Canada states that the effects of air density on the "inclined" trajectory are not relevant compared to the inclination effects of range and projection of velocity along the x-axis .

Thus, the departure angle needed to hit the target at the slant range is approximately

$$A_0 = A_S + \alpha_0.$$ (2.23.4)

The ballistics coefficient of the trajectory of flight corresponds to angle A_0. Using the ballistics coefficient law (2.23.3) we find that the BC of the projectile trajectory is

$$C_T = f(A_0).$$ (2.23.5)

The actual trajectory and the rigid trajectory are related by the relations (2.7.24)-(2.7.34) of the projectile-streamline and Snell's law model with the similitude factor (similarity parameter):

$$J = C_T / c_T,$$ (2.23.6)

Thus, the elements of both trajectories are related by the relations:

$$X_T = J \cdot x_T,$$ (2.23.7)

$$Y_T = J^2 \cdot y_T$$ (2.23.8)

$$T_T = J \cdot t_T$$ (2.23.9)

and

$$A_0 = ar\sin(J \cdot \sin\alpha_0).$$ (2.23.10)

Since the density temperature and pressure for similitude trajectories are considered to be equal, it is obvious that the parameter J, that relates two similitude trajectories with different ballistics coefficients, can be calculated using (2.23.6).

The actual point of impact of the projectile is the point with coordinates (2.23.7) and (2.23.8).

Because the ballistics coefficient of the projectile changes with the departure angle, the projectile will miss the target, and so we have to adjust the departure angle in order to hit the target at the actual inclined range.

When the angle of sighte is positive, projectile will pass over the target, and if the angle of sighte is negative, the projectile does not reach the target.

We have to reverse the problem in order to find for what departure angle the projectile will hit the target on the inclined range? (figure 7).

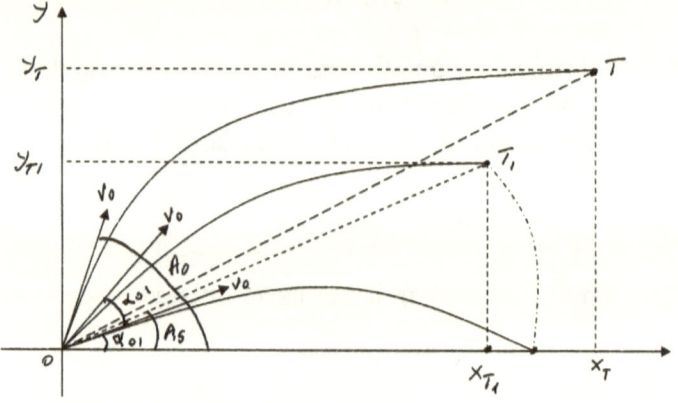

Figure 7

To make it easy to understand the use of the "general projectile streamline and Snell's law model" to determine the departure angle for the inclined fire, hereafter are shown two examples.

Example 23.1 Inclined Shooting

The target is located on a 15 degree inclined plane at a slant distance of 13252 meters, at the point with coordinates (12800m, 3430m).

Determine the projection angle needed to hit the target using the standard range table of 122mm Russian projectile. Projectile mass is 27.3kg, initial speed of 770m/s, mass of propellant charge is 6.53kg. The ballistics coefficient function (related with the G_{43}-function of resistance) is presented in the following table [52]:

---

52.   "Range Tables of 122mm cannon Model 1960 ", p. 117, Ministry of Defense of Albania, 1967

Table 1: The BC of 122mm projectile, departure speed 770m/s.

Angle, α_0	$\leq 5°$	$15°$	$25°$	$35°$	$45°$
BC, $c = c(\alpha_0)$	0.510	0.520	0.523	0.526	0.529

The departure angle of the projectile that corresponds to the horizontal range $x_H = 12800m$ is $\alpha_0 = 14.016667°$.

The other elements of the trajectory at the given horizontal range are: time of flight is 30s; terminal speed is 311m/s; the impact angle is 26 degrees.

Solution:

Using table 1, by interpolation, we find that the ballistics coefficient that corresponds to $\alpha_0 = 14.016667°$ is $c_T = 0.51902$.

As a first approach, using the principle of rigidity of the trajectory, we assume that the projection angle is equal to the departure angle $\alpha_0 = 14.016667°$.

Firing at the estimated angle the projectile will pass over the target and fall into another point.

The ballistics coefficient of the surpassing trajectory corresponds to the departure angle:

$$A_0 = A_S + \alpha_0 = 15° + 14.016667° = 29.016667°$$

From table 1, we find the corresponding ballistics coefficient:

$$C_T = f(29.01667°) = 0.52421.$$

Thus,

$$J = \frac{C_T}{c_T} = \frac{0.52421}{0.51902} = 1.01.$$

The projectile falls at the point with coordinates

$$X_T = J \cdot x_T = (1.01)(12800) = 12928,$$

and

$$Y_T = J^2 y_T = (1.01)^2 \cdot 3252 = 3317 .$$

It is obvious that the projectile surpasses the target.

Let's reverse the process. In order that the projectile should reach the point with coordinates (12800, 3250) the rigid trajectory should correspond to the point with coordinates

$$X_{T1} = J^{-1} \cdot x_T = (1.01)^{-1}(12800) = 12673$$

and

$$Y_{T1} = J^{-2} y_T = (1.01)^{-2} \cdot 3252 = 3188 .$$

The ballistics coefficient of the rigid trajectory 1 will be the ballistics coefficient that corresponds to the departure angle that corresponds to the horizontal range $X_{T1} = 12673$. From the standard range table of the given projectile, we find that the departure angle that corresponds to $X_{T1} = 12673$ is

$$\alpha_{01} = 13.7415° .$$

The corresponding departure angle is

$$A_{01} = A_S + \alpha_{01} = 15° + 13.7415° = 28.7415° .$$

Let's find the new value of parameter J.
Using table 1, we find that the ballistics coefficients are respectively:

$$c_{T1} = f(\alpha_{01}) = f(13.7415°) = 0.51874$$

and

$$C_{T1} = f(A_{01}) = f(28.7415°) = 0.524112 .$$

We find as well:

$$J = \frac{C_T}{c_T} = \frac{0.52412}{0.51874} = 1.010373$$

The similitude departure angle to hit the target on the inclined plane is

$$A_0 = \arcsin(J \cdot \sin A_{01}) = \arcsin(1.010373 \cdot \sin(28.7415)) = 29.068°.$$

The terminal point of impact of the projectile is the point with coordinates

$$x_T = J \cdot X_{T1} = (1.010373)(12673) = 12804$$

and

$$y_T = J^2 Y_{T1} = (1.010373)^2 \cdot 3188 = 3254.50.$$

Verification:

Using the PC program RPONIT.BAS for the following data input:

> departure angle, $A_0 = 29.068°$; ballistics coefficient, $C_{T1} = 0.524112$; x-coordinate of a point on trajectory, $x_T = 12804$, etc., ., we find that the y-coordinate of the point of impact is 3,252m.

Example 23.2

The target is located on a 10-degree inclined plane at a slant distance of 10357.4 meters, at the point with coordinates (10200m, 1798.54m).

Determine the projection angle needed to hit the target using the standard range table of 122mm Russian projectile, with a mass of 27.3kg, initial speed of 770m/s, and propellant charge mass of 6.53kg and the function of ballistics coefficient (related with the G_{43}-function of resistance) presented in the table 1 of example 23.1.

Use the PC program APROJC.BAS to find the elements of the trajectory for any horizontal range, we might need for the calculation of the inclined fire.

Solution:

Using APROJC.BAS, we find that the departure angle for the given projectile for the horizontal range of 10,200 meters is $\alpha_0 = 9.1157°$ (ballistics coefficient 0.51405 is obtained using an initial ballistics coefficient 0.515, and then using the table of ballistics coefficients to find the right value).

As in example 22.1, we find:

$$A_0 = A_S + \alpha_0 = 10° + 9.1157° = 19.1157°.$$

Using table of ballistics coefficients we find:

$$C_{T1} = f(A_{01}) = f(19.1157°) = 0.52123.$$

We have:

$$J = \frac{C_T}{c_T} = \frac{0.52123}{0.51405} = 1.01398.$$

In order that the projectile should reach the target located at the point with coordinates (10200, 1798.54) the rigid trajectory should correspond to the impact point with coordinates

$$X_{T1} = J^{-1} \cdot x_T = (1.01398)^{-1}(10200) = 10059.40$$

and

$$Y_{T1} = J^{-2} y_T = (1.01398)^{-2} \cdot 1798.54 = 1749.30.$$

The ballistics coefficient of the rigid trajectory 1 will be the ballistics coefficient that corresponds to the departure angle that corresponds to the horizontal range $X_{T1} = 10059.04$. Using again the APROJC.BAS, we find that the departure angle that corresponds to $X_{T1} = 10059.04$ ($c_{T1} = 0.5139$) is

$$\alpha_{01} = 8.8955°.$$

The corresponding departure angle is

$$A_{01} = A_S + \alpha_{01} = 10° + 8.8955° = 18.8955° \, .$$

Let's find the new value of parameter J.
Using table 1, we find that the ballistics coefficient that corresponds to $A_{01} = 18.8955°$ is:

$$C_{T1} = f(18.8955°) = 0.52117 \, .$$

We find as well:

$$J = \frac{C_T}{c_T} = \frac{0.52117}{0.5139} = 1.01414$$

The similitude departure angle is

$$A_0 = \arcsin(J \cdot \sin A_{01}) = \arcsin(1.01414 \cdot \sin(18.8955) = 19.17312° \, .$$

The impact of the projectile will be at the point with coordinates

$$x_T = J \cdot X_{T1} = (1.01414)(10059.04) = 10201.32$$

and

$$y_T = J^2 Y_{T1} = (1.01414)^2 \cdot 1749.30 = 1774 \, .$$

Note:

Using RPOINT.BAS, we find that the y-coordinate of the impact point is 1764.

3

APPROXIMATE ANALYTICAL SOLUTIONS

Introduction

In this chapter, we introduce the Siacci's method and some approximate analytical solutions of the system of differential equations of the projectile motion based on the analytical G-functions of resistance, introduced in chapter 1.

The Siacci's method is still used mainly in exterior ballistics of small arms.

To solve the exterior ballistics problems using Siacci's methods, the ballisticians use the prepared tables of four "Primary Siacci's Functions" that correspond to a given G-drag function. [53].

The Siacci's methods, based on the analytical G-functions, allow us to obtain approximate solutions and the equation of trajectory of projectile flight more or less in analytical form, avoiding the use of tabulated primary functions.

The Siacci's method can be used to solve the main problems of exterior ballistics and to construct the range tables.

For relatively small firearm calibers, the Siacci's method can be used to find the ballistics coefficient function and

---

[53.] McCoy, Robert, "Modern Exterior Ballistics", section 6.4, p. 100, Schiffer publishing 1999.

construct the range tables for departure angle not greater than 25-30 degrees. [54]

In chapter 4, we apply the Siacci's solution method to find range corrections of shooting.

For some special cases, in this chapter, we expose some other methods to solve the system of differential equations of projectile flight to obtain approximate analytical solutions and the equation of the projectile trajectory (for the projectile speed of less than 256 m/s).

To solve the ballistics problems for projectile speed less than 256 m/s, the reader can use the results we have obtained in *"Exterior Ballistics with Applications"*, chapter 4, adapting the results already obtained for the Siacci's approximate function of resistance.

The Siacci's method and the approximate analytical formulas we present here represent an expansion of the formulas that describe the projectile flight already presented "in chapter 3, and chapter 5 of *"Exterior Ballistics with Applications.*

3.1. Siacci's Method for High-Speed Projectiles

The Siacci's method of solution of the system of differential equations of projectile flight (2.1.3) gives acceptable results for departure angles not greater than 15 degrees.

The Siacci's method for the solution of the differential equations of projectile flight is introduced in chapter 5 of *"Exterior Ballistics with Applications"*. It is based on the approximate Siacci's G-function (1.6.6),

$$K_D(v) = \begin{cases} 1.212 \cdot 10^{-4} v^2 & for \quad v \le 256 m/s \\ (v-240)/3 & for \quad v > 256 m/s \end{cases}.$$

[54.] Shapiro, J. M., Exterior Ballistics", p. 278, Oborongiz, Moscow 50'.

The reader can find more details on the Siacci's solution method involving the use of the $K_D(v)$ drag function the reader can find in the book "*Exterior Ballistics with Applications*".

The reader should refer as well to the "*Exterior Ballistics of Small Arms*" to solve a variety of ballistics problems of small arms fire, using the rifleman's rules, the Siacci's trajectories, and other techniques presented in that book.

Differential Equations of Projectile Flight in Air

We will generalize the Siacci's solution method for all analytical G-functions of resistance given in sections 1.4-1.6 of the present book by formally replacing the approximate Siacci's G-function $K_D(v)$ with any G-drag function that is shown in sections 1.4-1.6.

Consider the vector differential equation of the projectile flight (1.7.2),

$$\frac{d\vec{v}}{dt} = \vec{g} - c \cdot h(y) \cdot G_D(v)\frac{\vec{v}}{v},$$

and the system of differential equations (2.1.4),

$$\begin{cases} \dfrac{dv_x}{dx} = -c \cdot h(y) \cdot \dfrac{G_D(v)}{v} \\ \dfrac{dp}{dx} = -\dfrac{g}{v_x^2} \\ \dfrac{dt}{dx} = \dfrac{1}{v_x} \qquad (p = \tan\alpha), \\ \dfrac{dy}{dx} = p \end{cases} \qquad (3.1.1)$$

where

$$G_D(v) = \begin{cases} A \cdot v^2 & for \quad v \le 256m/s \\ E \cdot v - F & for \quad v > 256m/s \end{cases}.$$

The values of A, E, and F are given in section (1.4), (1.5), and (1.6) of the book.

The G-function can be written in the following form:

$$G_D(v) = \begin{cases} A \cdot v^2 & for \quad v \le 256m/s \\ E(v - F/E) & for \quad v > 256m/s \end{cases} \qquad (3.1.2)$$

this is similar to the form of the approximate Siacci's function of resistance $K_D(v)$.

The Siacci's solution method of the differential equations (3.1.1) is based on the following assumptions:

- The departure angle is not greater than 15 degrees.
- The air density is the same at any point on the projectile trajectory. In other words, we assume that the density function is constant and equal to an average value $h(\bar{y})$, defined hereafter, i.e., we assume that $h(y) = h(\bar{y})$.

To solve the differential equations (3.1.1), Francisco Siacci introduced the projectile pseudo-speed u defined by the equation:

$$u = v \frac{\cos \alpha}{\cos \alpha_0}.$$

Following Siacci's method, we will solve the differential equations of projectile flight for any G-drag function, first for high-speed projectiles, i.e., when the projectile speed during the flight to the target is greater than 256m/s, and latter for slow-speed projectiles launched with initial speed of less than 256m/s.

We consider a standard atmosphere (ASM, ICAO, or TSM) assuming that density does not change with altitude.

Using (3.1.1) and the same procedures (and assumptions) as in sections 5.1 and 5.2 of the book "*Exterior Ballistics with Applications*", we can write the systems of differential equations of projectile flight (3.1.1) [55] as follows:

---

55. Klimi, G., "Exterior Ballistics with Applications", p.225, Xlibris, 2008

$$\begin{cases} \dfrac{dp}{du} = \dfrac{1}{B(u - F/E) \cdot u \cdot \cos^2 \alpha_0} \\[2mm] \dfrac{dt}{du} = -\dfrac{1}{gB(u - F/E) \cdot \cos \alpha_0} \\[2mm] \dfrac{dx}{du} = -\dfrac{u}{gB(u - F/E)} \\[2mm] \dfrac{dy}{du} = -p\dfrac{u}{gB(u - F/E)} \end{cases} \qquad (3.1.3)$$

where

$$p = v_y / v_x = \tan\alpha, \qquad u = v\dfrac{\cos\alpha}{\cos\alpha_0} \quad \text{(pseudo-speed)}, \qquad (3.1.4)$$

$$c = \dfrac{id^2}{m}1000, \qquad b = E \cdot c / g, \qquad B = \beta \cdot b. \qquad (3.1.5)$$

For anti-aircraft fire (or uphill/downhill fire):

$$\beta = h(\bar{y}) / \cos\alpha_0. \qquad (3.1.6)$$

For field—artillery fire (the muzzle of the gun and target are at the same horizontal firing site):

$$\beta = \dfrac{h(\bar{y})}{\sqrt{\cos\alpha_0}}. \qquad (3.1.7)$$

The density function for the flight of the projectile in the standard atmosphere is

$$h(\bar{y}) = (\dfrac{\tau_{0N} - 0.006328\bar{y}}{\tau_{0N}})^{4.4},$$

where:

- $\bar{y} = (2/3)y_m$, for field artillery, target on the ground,
- $\bar{y} = (1/2)y_m$, for anti-aircraft artillery (uphill fire); target is on the ascending point of the trajectory,; y_m is the maximum altitude of the projectile trajectory.

- τ_{0N} is the virtual temperature at the firing site at the sea level.

To make simpler the study of the trajectory of flight using Siacci's method, we can use the following approximations:

- According to Alger, the coefficient β, given by (3.1.7) and (3.1.8), can be considered equal to one,

$$\beta = 1,$$

if the time of the projectile flight to the target is less than 12 seconds. (This condition is satisfied for all small arms.).

- When the departure angle is less than 5 degrees. The pseudo-speed of the projectile u can be considered equal to the speed of the projectile v.[56]
- The coordinates (x_m, y_m) of the trajectory vertex can be estimated approximately using the formulae [57]:

$$x_m = (0.5 + 0.0001 \cdot v_0) \cdot x_T ,$$
$$y_m = 0.25 x_T [(0.5 + 0.0001 \cdot v_0) \cdot \tan(\alpha_0) + (0.5 - 0.0001 \cdot v_0) |\tan \alpha_T |].$$
$$(3.1.8)$$

where x_T and α_T are respectively the horizontal range of fire and the terminal angle.

- The ordinate of the trajectory vertex can be estimated with a good accuracy employing the formula [58]:

$$y_m = \frac{g \cdot t_T^2}{8} ,\qquad\qquad (3.1.9)$$

[56.] Alger, R. Philip, "Exterior Ballistics", p. 53, The Lord Baltimore Press, 1906.
[57.] Okunev, B. H "Fundamentals of Ballistics", Vol.1, Book 2, p. 186, Moscow, 1943
[58.] Alger, R. Philip, "Exterior Ballistics", p. 34, p. 60, The Lord Baltimore Press, 1906.

where t_m is the time of flight to the point of impact on the ground for the projectile flying in presence of drag.

The last approximation is very useful for the study of the trajectories of small arms projectiles. As it is shown in example 3.1, we see that formula (3.1.8) and (3.1.9) give acceptable results.

Other approximate formulae to estimate the coordinates of the trajectory vertex of a projectile are [59]:

$$x_m = 0.25 \cdot x_T + y_m \cdot \cot \alpha_0, \qquad y_m = 0.25 \cdot x_T \sqrt{\tan \alpha_0 \cdot \tan \alpha_T} \ .$$

Siacci's Formulae

The solution of the differential equations (3.1.3) gives the elements of the trajectory of flight in the following form: [60]:

The abscissa of a point on the trajectory is

$$x = -\frac{1}{Bg}[D(u) - D(v_0)], \qquad (3.1.10)$$

where:

$$D(u) = u + (F / E) \cdot \ln(u - F / E),$$
$$D(v_0) = v_0 + F / E \cdot \ln(v_0 - F / E). \qquad (3.1.11)$$

The corresponding *ordinate* of the point with abscissa given by (3.1.10) is

$$y = p_0 x + \frac{x}{(F / E) B \cos^2 \alpha_0} \cdot [\frac{A(u) - A(v_0)}{D(u) - D(v_0)} - J(v_0)], \qquad (3.1.12)$$

---

59. Cranz, C., Becker, K., Exterior Ballistics, p.244, London, 1921
60. The solution of the differential equations (3.1.3) can be obtained following the solution method presented in section 5.2 of the "Exterior Ballistics with Applications", Xlibris, 2008.

where

$$A(u) - A(v_0) = \int_{v_0}^{u} \frac{u}{u - F/E} \cdot \ln \frac{u - F/E}{u} \, du. \qquad (3.1.13)$$

Time of flight to a point with coordinates (x, y) given by (3.1.10) and (3.1.12) is

$$t = -\frac{1}{Bg \cos \alpha_0} [T(u) - T(v_0)], \qquad (3.1.14)$$

where

$$T(u) = \ln(u - F/E), \text{ and } T(v_0) = \ln(v_0 - F/E). \qquad (3.1.15)$$

The angle of flight at the trajectory point (x, y) is

$$p = p_0 + \frac{1}{(F/E)B \cos^2 \alpha_0} [J(u) - J(v_0)], \qquad (3.1.16)$$

where

$$J(u) = \ln(\frac{u - F/E}{u}), \quad J(v_0) = \ln(\frac{v_0 - F/E}{v_0}) \qquad (3.1.17)$$

and

$$p = \tan \alpha, \quad p_0 = \tan \alpha_0, \quad (\alpha \text{ is the angle of flight at (x, y)}).$$

The *departure angle* needed to hit the target located at the horizontal range x_T, at the point on the trajectory with coordinates $(x_T, 0)$ is given by the equation:

$$\sin(2\alpha_0) = -\frac{2}{(F/E)B} \cdot [\frac{A(u) - A(v_0)}{D(u) - D(v_0)} - J(v_0)]. \qquad (3.1.18)$$

The departure angle needed to hit the target located on the trajectory at the point with coordinates (x_T, y_T) is given by the equation:

$$\tan \alpha_0 + \frac{1}{(F/E)B\cos^2 \alpha_0} \cdot [\frac{A(u) - A(v_0)}{D(u) - D(v_0)} - J(v_0)] = \frac{y_T}{x_T}. \quad (3.1.19)$$

Formulae (3.1.18) and (3.1.19) are obtained from the equation of the projectile trajectory (3.1.12). Formula (3.1.9) is obtained assuming that the ballistics coefficient remains constant with increase/decrease of the departure angle (for angles till around 15 degrees).

The angle of site (inclined angle) A is given by the equation

$$\tan A = y_T / x_T.$$

Ballistics Coefficient

We assume that in practice of shooting, we are able to measure the angle of departure, the horizontal range, and the time of flight to the target. We consider that the projectile is fired on the ground at the sea level in standard atmosphere. The target is also located at the sea level.

The ballistics coefficient of the projectile can be determined by solving the system of equations (3.1.10) and (3.1.14), i.e.,

$$x = -\frac{1}{Bg}[D(u) - D(v_0)], \quad (3.1.20)$$

and

$$t = -\frac{1}{Bg\cos\alpha_0}[T(u) - T(v_0)], \quad (3.1.21)$$

where

$$B = \frac{c \cdot E \cdot h(\bar{y})}{g\sqrt{\cos\alpha_0}}, \quad h(\bar{y}) = (\frac{289.08 - 0.006328\bar{y}}{289.08})^{4.4}, \quad \bar{y} = 2y_{max}/3,$$

$$(3.1.22)$$

$$D(u) = u + (F / E) \cdot \ln(u - F / E),$$
$$D(v_0) = v_0 + F / E \cdot \ln(v_0 - F / E),   \quad (3.1.24)$$

and

$$T(u) = \ln(u - F / E), \text{ and } \quad T(v_0) = \ln(v_0 - F / E). \quad (3.1.25)$$

We can find B by solving the system of equations (3.1.20 and (3.1.21).

Indeed, eliminating B from the equations (3.1.20) and (3.1.21), we can write:

$$\frac{T(u) - T(v_0)}{t \cdot \cos(\alpha_0)} x = D(u) - D(v_0). \quad (3.1.26)$$

Solving the above equation for u (employing a graphing calculator TI-83+), and then substituting the obtained value of u in the equation (3.1.20) or (3.1.22), we find B. Then, using the first equation of (3.1.22), it is easy to find the ballistics coefficient of the given projectile.

For small arms, the departure angle of a projectile normally is very narrow. So, in the first formula of (3.1.22), we can consider the cosine of the departure angle equal to one and the pseudo-speed equal to the speed of the projectile.

Thus, for narrow angles, the first formula of (3.1.22) can be written in the form:

$$B = c\frac{E \cdot h(\bar{y})}{g}. \quad (3.1.27)$$

Primary Siacci's Functions

The functions,

$$D(u) = u + (F / E) \cdot \ln(u - F / E), \quad (3.1.28)$$

$$A(u) - A(v_0) = \int_{v_0}^{u} \frac{u}{u - F / E} \cdot \ln \frac{u - F / E}{u} du, \qquad (3.1.29)$$

$$T(u) = \ln(u - F / E), \qquad (3.1.30)$$

and

$$J(u) = \ln(\frac{u - F / E}{u}), \qquad (3.1.31)$$

that are present in the Siacci's formulas, are called "Siacci's primary functions". The values of the Siacci's primary functions, for a given value of the pseudo speed u can be estimated using a graphing calculator.

Notes:

- For departure angles that are less than 5 degrees, all the formulae and the primary functions obtained in this section, can be simplified using Alger's assumption that the pseudo-speed u is approximately equal to the projectile speed, i.e., considering that $u = v$.
- The above formulae can be further simplified if we consider Alger's assumption that for the time of flight not greater than 12 seconds, the parameter β, determined by (3.1.6) and (3.1.7), can be considered approximately one, i.e., $\beta = 1$.
- The solution of the ballistics problems can be simplified as well if we calculate the maximum ordinate of the trajectory using the approximate formula (3.1.9).
- The above notes are valid as well for the projectiles launched with speeds of less than 256m/s.

Example 1.1 Maximum Height of the Trajectory

Use the approximate formulae (3.1.8) and (3.1.9) to estimate the maximum altitude of a 122mm projectile fired at the an altitude of 3,000m by a Russian 122mm cannon with initial speed of 885m/s at an angle of 10 degrees if the time of flight is 26 seconds, the horizontal range is 14,000 meters, and the terminal angle is -18 degrees.

Solution:

Substituting in the second equation of (3.1.8), we find that the maximum altitude of the trajectory is

$y_m = 0.25(14000)[(0.5 + 0.0001 \cdot 885) \cdot \tan 10 + (0.5 - 0.0001 \cdot 885|\tan - 18|] = 831$

Using (3.1.9), we find

$$y_m = \frac{g \cdot t_T^2}{8} = \frac{g \cdot (26)^2}{8} = 829m.$$

The range table of the 122mm Russian cannon shows that the maximum altitude of the trajectory is 831 meters.

The abscissa of the trajectory vertex is approximately:

$x_m = (0.5 + 0.0001 \cdot v_0) \cdot x_T = (0.5 + 0.0001 \cdot 885) \cdot 14000 = 8239m.$

Example 1.2 Calculation of the Horizontal Range

A NATO 5.56×45mm M855 ball bullet is fired from an M16 rifle. The initial speed of the bullet is 945m/s, while the departure angle is 0.25073 degree. The ballistics coefficient relative to the G_7-function is $c = 9.4185m^3 / kg$ ($C = 0.151lb / in^2$). Assume that the projectile flight is in the ASM atmosphere at the sea level.

Find the range and all the other elements at the impact point considering G_7-function for the ASM atmosphere if the temperature and pressure at the sea level are standard (respectively 15 degrees Celsius and 750mm).

Solution:

The G-$_7$ function (see section 1.4) for the ASM atmosphere is

$$G_7(v) = \begin{cases} 5.66480 \times 10^{-5} \cdot v^2 & for & v \le 256m/s \\ 0.150355 \cdot v - 34.7319 & for & 256m/s < v \le 1700 \end{cases}.$$

We can write:

$$E = 0.150355, \quad F = 34.7319, \text{ and } \quad F/E = 231$$

Since the trajectory of the bullet is close to the ground, we can consider

$$\beta = h(\bar{y})/\cos\alpha_0 = 1,$$

$$b = E \cdot c/g = (0.150355) \cdot (9.4185)/9.80665 = 0.1444,$$

and

$$B = b\beta = 0.1444.$$

The equation (3.1.18) can be written:

$$\sin(2\alpha_0) = -\frac{1}{16.6782} \cdot [\frac{A(u) - A(v_0)}{D(u) - D(v_0)} - J(v_0)] \tag{1}$$

where

$$J(v_0) = \ln(\frac{v_0 - F/E}{v_0}) = \ln(\frac{945 - 231}{945}) = -0.2803,$$

and

$$D(v_0) = v_0 + F/E \cdot \ln(v_0 - F/E) = 2462.874.$$

Substituting $\alpha_0 = 0.25073$ in equation (1), we obtain

$$A(u) - A(v_0) = 0.1343326 \cdot [D(u) - D(v_0)] \tag{2}$$

where

$$A(u) - A(v_0) = \int_{945}^{u} \frac{u}{u - 231} \cdot \ln\frac{u - 231}{u} du$$

and
$$D(u) - D(v_0) = u + 231 \cdot \ln(u - 231) - 2462.874 .\qquad (3)$$

Using a graphing calculator TI83+ and a trial- and- error procedure from equation (2), we find that the pseudo-speed is

$$u = 480m/s .$$

Since the departure angle and the impact angle are relatively small, the pseudo-speed is approximately equal to the impact speed of the bullet, i.e., $v_T = 480m/s$.

The horizontal range is

$$x_T = -\frac{1}{Bg}[D(u) - D(v_0)] = -\frac{1}{0.1444g}[1754.5316 - 2462.874] = 500.20m$$

Using (3.1.14) and (3.1.15), we find that the time of flight is

$$t_T = -\frac{1}{0.1444g\cos(0.25075)}[T(480) - T(945)] = 0.744s .$$

The Trajectory Vertex

At the point of maximum altitude of the trajectory, the angle of flight is zero. Substituting $p = \tan\alpha = \tan(0) = 0$ in (3.1.16) and considering (3.1.17), we can write:

$$\tan(0.25075) + \frac{1}{(231)(0.1444)\cos^2(0.25075)}[J(u) - J(945)] = 0 \qquad (4)$$

where

$$J(945) = -0.2803, \text{ and } J(u) = \ln(\frac{u - 231}{u}) .$$

The equation (4) can be written

$$\ln(\frac{u - 231}{u}) = -.42628 .$$

Hence we find that the pseudo-speed of the bullet at the vertex is

$$u_m = 665.6 m / s.$$

Substituting the pseudo-speed in (3.1.10) and (3.1.12), we find that the coordinates of the trajectory vertex are respectively

$$x_m = -\frac{1}{Bg}[D(665.6) - D(945)] = -\frac{1}{0.1444g}[2068.8 - 2462.9] = 278.3m$$

and

$$y_m = \tan(0.25075)(278.3) + 8.3434 \cdot [\frac{A(665.6) - A(945)}{D(665.6) - D(945)} - J(945)] = 0.68m \cdot$$

For the maximum trajectory altitude, we obtain the same result using (3.1.9).
Indeed,

$$y_m = \frac{g \cdot t^2}{8} = \frac{g \cdot (0.744)^2}{8} = 0.68m.$$

Example 1.3 Calculation of the Departure Angle

A NATO $5.56 \times 45mm$ M855 ball bullet is fired from an M16 rifle. The initial speed of the bullet is 945m/s.

Find the departure angle needed to hit a target located at the horizontal range of 800 meters.

The firearm and the target are at the sea level. The ballistics coefficient relative to the G_1-function is $c = 5.0051m^2 / kg$
($C = 0.2842 lb / in^2$).

Find as well the time of flight.

Assume that the projectile flight is in the ASM atmosphere.

Solution:

The G_1-function for the ASM atmosphere is

$$G_1(v) = \begin{cases} 1.00347 \times 10^{-4} \cdot v^2 & for \quad v \le 256m / s \\ 0.312914v - 79.3976 & for \quad 256 < v \le 1000 \end{cases}$$

We can write:

$$E = 0.312914, \quad F = 79.3976, \text{ and } \quad F/E = 253.74$$

Since the trajectory of the bullet is close to the ground, we can consider

$$\beta = h(\bar{y})/\cos\alpha_0 = 1,$$

$$b = E \cdot c/g = (0.312914) \cdot (5.0051)/9.80665 = 0.1597,$$

and

$$B = b\beta = 0.1597.$$

The Pseudo-Speed at the Impact Point

The horizontal range is

$$x_T = -\frac{1}{Bg}[D(u) - D(v_0)] = -\frac{1}{0.1597g}[D(u) - D(945)] \qquad (1)$$

Substituting in (1):

$$x = 800, \text{ and}$$
$$D(945) = 945 + 253.74 \cdot \ln(945 - 253.74) = 2604.08,$$

we find that:

$$D(u) = 1351.90,$$

where

$$D(u) = u + 253.74 \cdot \ln(u - 253.74).$$

Thus, we can write

$$u + 253.74 \cdot \ln(u - 253.74) - 1351.90 = 0$$

Solving the last equation with a graphing calculator TI83+, we find that the pseudo-speed at the horizontal range $x = 800$ is

$$u_T = 313.6 \ m/s.$$

Since the departure angle is small, we can consider that the terminal speed of projectile at the impact point is 313.6m/s.

Departure angle

Using (3.1.18), for the departure angle of the projectile trajectory, we have:

$$\sin(2\alpha_0) = -\frac{2}{(253.74)(0.1597)} \cdot [\frac{A(313.6) - A(945)}{D(313.6) - D(945)} - J(945)], \qquad (2)$$

Substituting in (2):

$$A(u) - A(v_0) = \int_{945}^{313.6} \frac{u}{u - 253.74} \cdot \ln\frac{u - 253.74}{u} du = 936.52, \qquad (3)$$

(The integral in (3) can be calculated using a graphing calculator TI83+.)

We have:

$$D(313.6) - D(945) = -1252.90,$$

and

$$J(945) = \ln(\frac{v_0 - F/E}{v_0}) = \ln(\frac{945 - 253.74}{945}) = -0.312669.$$

We find that

$$\sin(2\alpha_0) = 0.2146. \qquad (4)$$

Hence, we find that the departure angle is

$$\alpha_0 = (\sin^{-1} 0.2146)/2 = 0.61483°.$$

The Time of Flight

Substituting in (3.1.14), for the time of flight to the target, we find:

$$t_T = -\frac{1}{Bg\cos\alpha_0}[T(u) - T(v_0)] = -\frac{1}{(0.1597)g}[T(313.6) - T(945)] = 1.562s\,,$$

since:

$$T(945) = \ln(945 - 253.74) = 6.5385\,,$$
$$T(313.6) = \ln(313.6 - 253.74) = 4.092\,.$$

The Maximum Ordinate of the Trajectory

Using (3.1.9), we find that the maximum altitude of the trajectory is

$$y_m = \frac{g \cdot t^2}{8} = \frac{g \cdot (1.562)^2}{8} = 2.99m\,.$$

Example 1.4 Departure Angle, Construction of Range Tables

The ballistics coefficient estimated experimentally is a function of the departure angle α_0, i.e. $c = f(\alpha_0)$.

Using the firing tests for a set of angeles (in degree) 5, 10, 15, 25, 35, 45, the ballisticians determine the horizontal range, and then the respective ballistics coefficients.

Once the table of ballistics coefficients is constructed, we have an approximate function $c = f(\alpha_0)$ of ballistics coefficient in form of a table. Using $c = f(\alpha_0)$, we are able to solve the differential equations of projectile flight, find the elements of the trajectory for any horizontal range, and construct the range tables.

The example presented hereafter demonstrates the use of Siacci's method to find the elements of the trajectory for a given horizontal range in standard atmosphere.

Data: The ballistics coefficient related with the G-function of 1943, for the Russian 122mm-caliber field cannon caliber 122mm fired at the initial speed of 885m/s are obtained using the following table: [61]

Table 1: The BC of 122mm Projectile, Departure Speed 885m/s.

Angle, α_0	$\leq 5°$	$15°$	$25°$	$35°$	$45°$
BC, $c=f(\alpha_0)$	0.496	0.514	0.516	0.519	0.521

Find the departure angle needed to hit the target located at the horizontal range $x_T = 11000m$ in TSA atmosphere, as well as the other elements of the trajectory.

Solution:

First Approach

For the G_{43}-Function (section 1.6), we have:

$$E = 0.157713, \quad F = 36.39542, \text{ and } \quad F/E = 230.77.$$

As a first approach, we can consider

$$\beta = h(\bar{y})/\sqrt{\cos\alpha_0} = 1,$$

$$b = E \cdot c / g = (0.157713) \cdot (0.5) / 9.80665 = 0.00804,$$

and

$$B = b\beta = 0.00804.$$

The Pseudo-Speed at the Terminal Point

The horizontal range is

$$x_T = -\frac{1}{Bg}[D(u) - D(v_0)] = -\frac{1}{0.00804g}[D(u) - D(885)] \qquad (1)$$

---

61. "Range Tables of 122mm cannon Model 1960", p. 117, Ministry of Defense of Albania, 1967

Substituting in (1):

$x_T = 11000m$, and $D(885) = 885 + 230.77 \cdot \ln(885 - 230.77) = 2381.19$,

we find that:

$$D(u) = 1513.89,$$

where

$$D(u) = u + 230.77 \cdot \ln(u - 230.77).$$

Thus, we can write

$$u + 230.77 \cdot \ln(u - 230.77) - 1513.89 = 0$$

Solving the last equation with a graphing calculator TI83+, we find that the pseudo-speed at the horizontal range $x_T = 11000m$ is

$$u_T = 371.81 \ m/s.$$

The Departure Angle

Using (3.1.18), for the departure angle of the projectile, we have:

$$\sin(2\alpha_0) = -\frac{2}{(230.77)(0.00804)} \cdot [\frac{A(371.81) - A(885)}{D(371.81) - D(885)} - J(885)], \qquad (2)$$

Substituting in (2):

$$A(u) - A(v_0) = \int_{885}^{371.81} \frac{u}{u - 230.77} \cdot \ln\frac{u - 230.77}{u} du = 469.62, \qquad (3)$$

$$D(371.81) - D(885) = -867.3,$$

and

$$J(885) = \ln(\frac{v_0 - F/E}{v_0}) = \ln(\frac{885 - 230.77}{885}) = -0.30213,$$

we find that

$$\sin(2\alpha_0) = 0.2580. \qquad (4)$$

Hence, we find that the departure angle is

$$\alpha_0 = (\sin^{-1} 0.2580) / 2 = 7.4756°.$$

The Time of Flight

Substituting in (3.1.14), for the time of flight to the target, we find:

$$t_T = -\frac{1}{Bg\cos\alpha_0}[T(u) - T(v_0)] = -\frac{1}{0.00804g}[T(371.81) - T(885)] = 19.46s,$$

since

$$T(885) = \ln(885 - 230.77) = 6.4835,$$
$$T(371.81) = \ln(371.81 - 230.77) = 4.9490.$$

The Approximate Altitude of the Trajectory

The maximum altitude of the projectile trajectory is

$$y_m = \frac{g \cdot t^2}{8} = \frac{g \cdot (19.46)^2}{8} = 464.3m.$$

Improving the Accuracy

The density function is

$$h(\bar{y}) = (\frac{289.08 - 0.006328(464.3) \cdot (2/3)}{289.08})^{4.4} = 0.97053.$$

The Ballistics Coefficient

Using the table of ballistics coefficients (table 1), by interpolation, we find a new ballistics coefficient, $c = 0.5005$.

We find as well

$$B = \frac{c \cdot E \cdot h(\bar{y})}{g\sqrt{\cos\alpha_0}} = \frac{0.5005 \cdot (0.157713) \cdot (0.97053)}{9.80665 \cdot \sqrt{\cos(7.4757}} = 0.007845.$$

The Pseudo-Speed at the Impact Point

The horizontal range is

$$x_T = -\frac{1}{Bg}[D(u) - D(v_0)] = -\frac{1}{0.007845 \cdot g}[D(u) - D(885)] \qquad (1)$$

Substituting in (1):

$$x_T = 11000m, \quad \text{and}$$
$$D(885) = 885 + 230.77 \cdot \ln(885 - 230.77) = 2381.19,$$

we find that:

$$D(u) = 1534.93,$$

where

$$D(u) = u + 230.77 \cdot \ln(u - 230.77).$$

Thus,

$$u + 230.77 \cdot \ln(u - 230.77) - 1534.93 = 0$$

Solving the last equation with a graphing calculator TI83+, we find that the pseudo speed at the horizontal range $x_T = 14000m$ is

$$u_T = 381.84 \ m/s.$$

The Departure Angle

Using (3.1.18), for the departure angle of the projectile, we can write:

$$\sin(2\alpha_0) = -\frac{2}{(230.77)(0.007845)} \cdot [\frac{A(381.84) - A(885)}{D(381.84) - D(885)} - J(885)], \qquad (2)$$

Substituting in (2):

$$A(u) - A(v_0) = \int_{885}^{381.84} \frac{u}{u - 230.77} \cdot \ln\frac{u - 230.77}{u} \, du = 445.06, \qquad (3)$$

$$D(381.84) - D(885) = -846.23$$

and

$$J(885) = \ln(\frac{v_0 - F/E}{v_0}) = \ln(\frac{885 - 230.77}{885}) = -0.30213,$$

we find that

$$\sin(2\alpha_0) = 0.2473. \tag{4}$$

Hence, we find that the departure angle is

$$\alpha_0 = (\sin^{-1} 0.2473) / 2 = 7.157°.$$

(Note that the integral in (3) is calculated using a graphing calculator TI83+.)

The Time of Flight

Substituting in (3.1.14), for the time of flight to the target, we find:

$$t_T = -\frac{1}{Bg\cos\alpha_0}[T(u) - T(v_0)] = -\frac{1}{0.007845g}[T(381.84) - T(885)] = 19.16s$$

,

since

$$T(885) = \ln(885 - 230.77) = 6.4835,$$
$$T(381.84) = \ln(381.84 - 230.77) = -0.3508.$$

The Approximate Altitude of the Trajectory

The maximum altitude of the trajectory is

$$y_m = \frac{g \cdot t^2}{8} = \frac{g \cdot (19.16)^2}{8} = 470m.$$

The Terminal Angle

The angle at the impact point is

$$\tan\alpha_T = \tan\alpha_0 + \frac{1}{(F/E)B\cos^2\alpha_0}[J(u) - J(v_0)] = \tan(7.4757) +$$

$$+ \frac{1}{(230.77)\cdot(0.007845)\cdot\cos^2(7.157)}[J(381.84) - J(885)] = -0.35076$$

Hence,

$$\alpha_0 = \arctan(-0.35076) = 20.5338°.$$

The Terminal Speed

The speed of the projectile at the impact point is

$$v_T = u \frac{\cos \alpha_0}{\cos \alpha} = 381.84 \frac{\cos(7.157)}{\cos(-20.5338)} = 405.$$

Example 1.5 Projectile Drop

(a) Use the results obtained in Example 1.2 to find the projectile drop at 800 meters if the bullet is launched horizontally (departure angel $\alpha_0 = 0°$).

Find as well the angle of sight α_s (the angle that the departure line forms with the line of sight) if the sight height is 0.0381meters (refer to section 1.5 of the book "*Exterior Ballistics of Small Arms*").

(b) Use the Siacci's formulas and the results of the above example to find the projectile drop at the horizontal range of 800m.

Solution:

(a) Approximate Direct Calculations

The departure angle that zeroes the firearm at the horizontal range $x = 800$ is $\alpha_0 = 0.61483°$. The projectile drop in absolute values is

$$Drop = x_T \cdot \tan \alpha_0 = (800) \cdot \tan(0.61483) = 8.58m .$$

Using the formulas (1.5.5) or (1.5.6) presented in the "*Exterior Ballistics of Small Arms*", i.e.,

$$\alpha_S = \alpha_0 - \tan^{-1}(h_s / x_T)$$

we find that the angle of sight is

$$\alpha_S = 0.61483 - \tan^{-1}(0.0381 / 800) = 0.6121° = 36.726 MOA .$$

(b) Siacci's Method

The equation of the trajectory (3.1.12) for the projectile of the example 1.3 when departure angle is zero ($\alpha_0 = 0°$; $p_0 = 0°$), is:

$$y = \frac{x}{(F/E)B} \cdot [\frac{A(u) - A(v_0)}{D(u) - D(v_0)} - J(v_0)] \,.$$

Using (3.1.10), the above equation can be written:

$$y = -\frac{1}{(F/E)B} \cdot [\frac{A(u) - A(v_0)}{Bg} - J(v_0)x] \,.$$

Substituting in (3.1.12) the values we have in example (1.3),:

$E = 0.312914$, $F = 79.3976$, and $F/E = 253.74$, $B = b\beta = 0.1597$, $u = 313.6m/s$,

for the point on the trajectory with abscissa $x = 800$, we find that the drop is

$$y = -\frac{1}{(253.74) \cdot (0.1597)} \cdot [\frac{A(313.6) - A(945)}{(0.1597)g} - J(945) \cdot (800)] = -8.505m \cdot$$

Note that for both methods, the calculated drop is approximately the same.

Example 1.6 Lapua Scenar GB528, 19.44 g, Bullet; G_1-Function

A Lapua Scenar GB528 19.44 g bullet is fired horizontally at the sea level with an initial speed of 830 m/s . The ballistics coefficient relative to the G_1-function of resistance is $C = 0.785$ lb/in^2 [62]; $c = 1.422/C = 1.8115$ m^2/kg .

---

62. Wikipedia Contributors, "External Ballistics", Wikipedia, The Free Ency-clopedia, http://en.wikipedia.org/wiki/External_ballistics (accessed October 24, 2009).

Use the Siacci's method to find *the projectile drop*, *the terminal speed*, and the *time of flight* of the given bullet at the horizontal ranges: 300, 600, 900, 1,200, 1,500, 1,800, and 2,400 meters.

The shooting is at the sea level in ICAO atmosphere.

Solution:

Hereafter, it is shown the solution only the solution for the horizontal range of 1,200 meters is shown.

The G_1-function of resistance for the projectile trajectory in ICAO atmosphere is

$$G_{1A}(v) = \begin{cases} 1.0584 \times 10^{-4} \cdot v^2 & for \quad v \le 256 m/s \\ 0.315754 \cdot v - 78.6769 & for \quad 256 < v \le 1000 \end{cases}. \quad (1)$$

For the G_{1A}-function, we have:

$$E = 0.315754, \; F = 78.6769, \text{ and } F/E = 249.1715.$$

Since the trajectory of the bullet is close to the ground, we can consider

$$\beta = h(\bar{y}) / \cos\alpha_0 = 1,$$

$$b = E \cdot c / g = (0.315754) \cdot (1.8115) / 9.80665 = 0.05833,$$

and

$$B = b\beta = 0.05833.$$

The equation of the trajectory (3.1.12) for when departure angle is zero ($\alpha_0 = 0°$; $p_0 = 0°$) is:

$$y = \frac{x}{(F/E)B} \cdot [\frac{A(u) - A(v_0)}{D(u) - D(v_0)} - J(v_0)].$$

Using (3.1.10), the above equation can be written:

$$y = -\frac{1}{(F/E)B} \cdot [\frac{A(u) - A(v_0)}{Bg} - J(v_0)x]. \quad (2)$$

The terminal speed at the horizontal range of 1200 meters

The horizontal range is

$$x_T = -\frac{1}{Bg}[D(u) - D(v_0)] = -\frac{1}{0.05833g}[D(u) - D(830)] \qquad (1)$$

Substituting in (1):

$x = 1200$, and
$D(830) = 830 + 249.1715 \cdot \ln(830 - 249.1715) = 2415.84$,

we find that:

$D(u) = 1729.41$,

where

$D(u) = u + 249.1715 \cdot \ln(u - 249.1715)$.

Thus, we can write

$$u + 249.1715 \cdot \ln(u - 249.1715) - 1729.41 = 0$$

Solving the last equation with a graphing calculator TI83+, we find that the pseudo- speed at the horizontal range $x = 1200$ is

$u_T = 431.83 \; m/s$.

Since the departure angle is small, we can consider that the terminal speed of projectile at the impact point is 431.83 m/s, i.e.,

$v_T = 431.83 \; m/s$.

The Projectile Drop

The projectile drop at the 1,200-meter horizontal range 1200 meters is

$$y = -\frac{1200}{(249.1715) \cdot (0.05833)} \cdot [\frac{A(431.83) - A(830)}{D(431.83) - D(830)} - J(830)] = -16.31m,$$

since

$$A(u) - A(v_0) = \int_{830}^{431.83} \frac{u}{u - 249.1715} \cdot \ln\frac{u - 249.1715}{u} \, du = 380.67 , \qquad (3)$$

and

$$J(830) = \ln(\frac{v_0 - F/E}{v_0}) = \ln(\frac{830 - 249.1715}{830}) = -0.35697 .$$

The Time of Flight

For the time of flight to the point with coordinates (1200, -16.31), we find:

$$t = -\frac{1}{Bg \cos\alpha_0}[T(u) - T(v_0)] = -\frac{1}{0.05833g}[T(431.83) - T(830)] = 2.022 \ s ,$$

since

$$T(431.83) = \ln(431.83 - 249.1715) = 5.2076 ,$$

and

$$T(830) = \ln(830 - 249.1715) = 6.36446 .$$

In the same way, it is estimated the projectile drop, the terminal speed, and the time of flight for the other horizontal distances are estimated. The obtained results are presented in table 1.

Table 1: Lapua Scenar GB528 19.44 g Bullet Drop—G_1 Function

Range	300	600	900	1200	1500	1800	2100
Speed	714	607	512	432	368	321	290
Time	0.39	0.845	1.384	2.022	2.777	3.653	4.639
Drop	- 0.709	- 3.166	- 8.043	- 16.31	- 29.36	- 48.94	- 77.07

Note:
The results obtained using a fixed ballistics coefficient of 1.8115 (especially for ranges greater than 600 meters) have relatively big errors.

That is why it is recommended to measure the ballistics coefficient for ranges of 500, 800, 1,500, 2,000, 3,000, 4,000 meters (see introduction, chapter 2).

Example 1.7 Lapua Scenar GB528 19.44 g, Bullet; G_7-function

A Lapua Scenar GB528 19.44 g bullet is fired horizontally at the sea level with an initial speed of 830 m/s. The ballistics coefficient [63] relative to the G_7-function of resistance is $C = 0.377$ lb / in^2;

$c = 1.422 / C = 3.7724$ m^2 / kg.

Use the Siacci's method to find the projectile drop, the terminal speed, and the time of flight of the given bullet at the horizontal ranges: of 300, 600, 900, 1,200, 1,500, 1,800, and 2,400 meters.

Shooting is at the sea level in ICAO atmosphere.
Solution:

The G_7-function of resistance for the projectile trajectory in ICAO atmosphere is

$$G_{7A}(v) = \begin{cases} 5.7679 \times 10^{-5} \cdot v^2 & for & v \le 256 m / s \\ 0.152593 \cdot v - 35.1717 & for & 256 < v \le 1700 m / s \end{cases}.$$

We have: $E = 0.152593$, $F = 35.1717$, and $F / E = 230.4935$.

Since the trajectory of the bullet is close to the ground, we can consider

$$\beta = h(\bar{y}) / \cos \alpha_0 = 1,$$

$$b = E \cdot c / g = (0.152593) \cdot (3.7724) / 9.80665 = 0.058699,$$
and

$$B = b\beta = 0.058699.$$

The terminal speed at the horizontal range of 1,200 meters

The horizontal range is

$$x_T = -\frac{1}{Bg}[D(u) - D(v_0)] = -\frac{1}{0.058699g}[D(u) - D(830)]. \qquad (1)$$

---

63. Wikipedia Contributors, "External Ballistics", Wikipedia, The Free Encyclopedia, http://en.wikipedia.org/wiki/External_ballistics (accessed October 24, 2009).

Substituting in (1):,

$x = 1200$, and
$D(830) = 830 + 230.4935 \cdot \ln(830 - 230.4935) = 2304.261$,

we find that:
$$D(u) = 1613.49,$$

where

$$D(u) = u + 230.4935 \cdot \ln(u - 230.4935).$$

Thus, we can write

$$u + 230.4935 \cdot \ln(u - 230.4935) - 1613.49 = 0$$

Solving the last equation with a graphing calculator TI83+, we find that the pseudo-speed at the horizontal range $x = 1200$ is

$$u_T = 413.16 \ m/s.$$

We can consider that the terminal speed of projectile at the impact point is 413.16 m/s, i.e., $v_T = 413.16 \ m/s$.

The Projectile Drop

Since the departure angle is small, we can consider that the terminal speed of projectile at the impact point is 431.83 m/s, i.e., $v_T = 431.83 \ m/s$.

The projectile drop is:

$$y = -\frac{1200}{(230.4935) \cdot (0.058699} \cdot [\frac{A(413.16) - A(830)}{D(413.16) - D(830)} - J(830)] = -16.779m$$

since

$$A(u) - A(v_0) = \int_{830}^{413.16} \frac{u}{u - 230.4935} \cdot \ln\frac{u - 230.4935}{u} du = 355.40, \quad (3)$$

and

$$J(830) = \ln(\frac{v_0 - F/E}{v_0}) = \ln(\frac{830 - 230.4935}{830}) = -0.32532.$$

Note that the integral in (3) can be calculated using a graphing calculator TI83+.

The Time of Flight

The time of flight to the point with coordinates (1200, -16.779) is

$$t = -\frac{1}{Bg\cos\alpha_0}[T(u) - T(v_0)] = -\frac{1}{0.058699g}[T(413.16) - T(830)] = 2.0646 \; s \text{,}$$

since

$$T(413.16) = \ln(413.16 - 230.4935) = 5.2077$$

and

$$T(830) = \ln(830 - 230.4935 = 6.3961.$$

In the same way, it is estimated the projectile drop, the terminal speed, and the time of flight for the other horizontal distances are estimated. The obtained results are presented in table 3.

Table 3: Lapua Scenar GB528 19.44 g Bullet Drop—G_7 Function

Range	300	600	900	1200	1500	1800	2100
Speed	709	598	498	413	346	298	267
Time	0.391	0.853	1.402	2.065	2.860	3.798	4.867
Drop	- 0.712	- 3.198	- 8.187	-16.78	- 30.59	- 51.81	- 82.96

3.2. Siacci's Method for Non-Standard Atmosphere

Traditionally, the Siacci's solution method is applied to solve the differential equations of projectile flight in standard atmosphere.

The characteristics of atmosphere change continuously not only during the week, month, or year but also even during a 24 hours day. The conditions of the standard atmosphere are rarely met in practice. The projectile is not always launched in an ideal standard atmosphere and at the sea level. For that reason, the range tables of a firearm usually contains corrections for small changes in the characteristics of the atmosphere.

The corrections for relatively small changes in the characteristics of the atmosphere are obtained using the theory of corrections. For example, we can use the theory of corrections, shown in "*Exterior Ballistics with Applications*", that is based on the Siacci's method.

The Siacci's theory of corrections (chapter 6, "*Exterior Ballistics with Applications*") is not accurate for large changes in the characteristics of the atmosphere.

Using the PC programs presented in chapter 2, we are able to calculate the elements of the projectile trajectory as well as the corrections when the projectile is launched in a non standard atmosphere.

Nevertheless, hereafter, we show the Siacci's methods to calculate the projectile trajectory of fast-flying projectiles in non-standard atmosphere, avoiding the application of the correction theory.

The formulae we obtain hereafter can be used in any shooting site, at the sea level or over the sea level, in mountain shooting, summer or winter shooting.

The projectile flight in non-standard atmosphere is described by the vector differential equation of projectile flight (2.6.9), i.e.,

$$\frac{d\vec{v}}{dt} = \vec{g} - \overline{c} \cdot h_2(y) \cdot G_D(v)\frac{\vec{v}}{v}, \qquad (3.2.1)$$

where

$$G_D(v) = \begin{cases} A \cdot v^2 & for \quad v \le 256m/s \\ E(v - F/E) & for \quad v > 256m/s \end{cases}. \qquad (3.2.2)$$

Consider as well the system of differential equations of projectile flight (2.6.12)

$$\begin{cases} \dfrac{dv_x}{dx} = -\bar{c} \cdot h_2(\bar{y}_2) \cdot \dfrac{G_D(v)}{v} \\[2mm] \dfrac{dp}{dx} = -\dfrac{g}{v_x^2} \\[2mm] \dfrac{dt}{dx} = \dfrac{1}{v_x} \\[2mm] \dfrac{dy}{dx} = p \end{cases} \qquad , \qquad (3.2.3)$$

where:

$$h_2(\bar{y}_2) = (\dfrac{\tau_2 - 0.006328\bar{y}_2}{\tau_2})^{4.4}, \quad \bar{c} = \dfrac{p_2}{p_{0N}}\sqrt{\dfrac{\tau_{0N}}{\tau_2}} \cdot c, \qquad (3.2.4)$$

and

$$\bar{y}_2 = (2/3)y_{m2}, \quad c = \dfrac{id^2}{m} 1000.$$

We denote ρ_{0N}, p_{0N}, and τ_{0N}, as respectively the density, pressure, and the virtual temperature of the given standard atmosphere (ASM, ICAO, TSA) and ρ_2, p_2, and τ_2 as the values of the same parameters but in the given non-standard atmosphere.

Consider the Siacci's assumptions for the projectile flight.
Since the form of differential equations (3.2.3) is the same as the form of differential equations (3.1.1), it is obvious that the Siacci's formulas for the non-standard atmosphere are the same as those obtained in section (3.1) except that in those formulas, we have to formally substitute the fictive ballistics coefficient $\bar{c}$ instead of the ballistics coefficient c, and $h_2(\bar{y}_2)$ instead of $h(\bar{y})$.

Thus, for the projectile flight in non standard atmosphere, we have the following Siacci's formulas:
The *x-coordinate* of a point on the trajectory is

$$x = -\dfrac{1}{Bg}[D(u) - D(v_0)], \qquad (3.2.5)$$

The corresponding *y-coordinate* of the point with abscissa (3.2.5) is

$$y = p_0 x + \frac{x}{(F/E)\overline{B}\cos^2\alpha_0} \cdot [\frac{A(u) - A(v_0)}{D(u) - D(v_0)} - J(v_0)], \qquad (3.2.6)$$

Time of flight to the point with coordinates (x, y) given by (3.2.5) and (3.2.6) is

$$t = -\frac{1}{\overline{B}g\cos\alpha_0}[T(u) - T(v_0)], \qquad (3.2.7)$$

The angle of flight at the point (x, y) is

$$p = p_0 + \frac{1}{(F/E)\overline{B}\cos^2\alpha_0}[J(u) - J(v_0)], \qquad (3.2.8)$$

where

$$\overline{c} = \frac{p_2}{p_{0N}}\sqrt{\frac{\tau_{0N}}{\tau_2}} \cdot c, \qquad c = \frac{id^2}{m}1000 \qquad (3.2.9)$$

For the field artillery,:

$$\overline{B} = \frac{\overline{c}\cdot E\cdot h_2(\overline{y}_2)}{g\sqrt{\cos\alpha_0}}, \qquad h_2(\overline{y}_2) = (\frac{\tau_2 - 0.006328\overline{y}_2}{\tau_2})^{4.4},$$

$$\overline{y}_2 = (2/3)y_{m2} \qquad (3.2.10)$$

For the inclined fire (anti-aircraft fire/downhill or uphill shooting),:

$$\overline{B} = \frac{\overline{c}\cdot E\cdot h_2(\overline{y}_2)}{g\cdot\cos\alpha_0}, \qquad h_2(\overline{y}_2) = (\frac{\tau_2 - 0.006328\overline{y}_2}{\tau_2})^{4.4}, \qquad (3.2.11)$$

$$\overline{y}_2 = y_{m2}/2.$$

The *departure angle* needed to hit the target located at the horizontal range x_T, i.e., at the point on the trajectory with coordinates $(x_T, 0)$, is given by the equation:

$$\sin(2\alpha_0) = -\frac{2}{(F/E)\overline{B}} \cdot [\frac{A(u) - A(v_0)}{D(u) - D(v_0)} - J(v_0)]. \qquad (3.2.12)$$

For the inclined fire, the departure angle needed to hit a target located at the point with coordinates (x_T, y_T) is given by the equation

$$\tan\alpha_0 + \frac{1}{(F/E)\overline{B}\cos^2\alpha_0} \cdot [\frac{A(u) - A(v_0)}{D(u) - D(v_0)} - J(v_0)] = \frac{y_T}{x_T}, \qquad (3.2.13)$$

where $A°$ is the angle of firing site given by the equation:

$$\tan A° = y_T / x_T. \qquad (3.2.14)$$

To use the Siacci's formulae (3.2.5)-(3.2.8), and (3.2.12)-, (3.2.13), we must calculate the values of the Siacci's primary functions

$$D(u) = u + (F/E) \cdot \ln(u - F/E), \qquad (3.2.15)$$

$$A(u) - A(v_0) = \int_{v_0}^{u} \frac{u}{u - F/E} \cdot \ln\frac{u - F/E}{u} du, \qquad (3.2.16)$$

$$T(u) = \ln(u - F/E), \qquad (3.2.17)$$

and

$$J(u) = \ln(\frac{u - F/E}{u}), \qquad (3.2.18)$$

for a given value of the pseudo-speed u:

$$u = v\frac{\cos\alpha}{\cos\alpha_0}. \qquad (3.2.19)$$

To make simpler the solution of the exterior ballistics problem, we can estimate the maximum height of the trajectory using (3.1.9), i.e.,

$$y_m = \frac{g \cdot t^2}{8} .$$ (3.2.20)

Example 2.1 Mountain Shooting

A 122mm artillery cannon projectile of mass $m = 27.30$ kg is fired with an initial speed of $v_0 = 885 m / s$. The firing site is 3,000 meters over the sea level. The temperature, the virtual temperature, and the pressure at the firing site are respectively: 5.508 degrees Celsius, 279.23 degrees Kelvin, and 625.90 mmHg. The atmosphere at the sea level is TSA (pressure, 750 mmHg; temperature, 15 degrees Celsius, relative humidity, 50% (pressure of water vapors is 6.35 mmHg).

The projectile ballistics coefficient relative to the G_2-function is $c = 0.5300$.

Determine the launching angle α_0 needed to hit a target located on the given site at the horizontal distance $x = 10,000 m$.

Solution:

The values of the constants E and F that correspond to G_2-function are respectively $E = 0.1413$ and, $F = 29.9097$ (formula 1.4.7), while $F / E = 211.675$.

Estimation of $\bar{B}$, using (3.2.10) and (3.2.9):

$$\bar{c} = \frac{p_2}{p_{0N}} \sqrt{\frac{\tau_{0N}}{\tau_2}} \cdot c = \frac{625.90}{750} \sqrt{\frac{289.08}{279.23}} (0.53) = 0.4500 m^2 / kg .$$

As a first approach, we consider $h_2(\bar{y}_2) = 1$ and $\cos \alpha_0 = 1$. So, we have:

$$\bar{B} = \frac{\bar{c} \cdot E \cdot h(\bar{y}_2)}{g \sqrt{\cos \alpha_0}} = \frac{(0.45002) \cdot (0.1413)}{g} = 0.0064842 .$$ (1)

The Pseudo-Speed at Impact Point

Using (3.2.5), we find

$$D(u_T) = D(v_0) - Bgx = D(885) - (0.0064842) \cdot g \cdot (10000), \qquad (2)$$

where

$$D(u_T) = u_T + (211.675) \cdot \ln(u_T - 211.675)$$

and

$$D(v_0) = 885 + (211.675) \cdot \ln(885 - 211.675) = 2263.48$$

Substituting in (2), we find,

$$D(u_T) = 1627.597,$$

and

$$u_T + (211.675) \cdot \ln(u_T - 211.675) = 1627.597. \qquad (3)$$

Solving equation (3) using a graphing calculator, we find the approximate value of the terminal pseudo-speed

$$u_T = 460.143.$$

The Departure Angle

Substituting in (3.2.12), we have:

$$\sin(2\alpha_0) = -\frac{2}{(211.675)(0.0064842)} \cdot [\frac{A(460.14) - A(885)}{D(460.14) - D(885)} - J(885)] = 0.19603 \cdot \qquad (4)$$

since:

$$A(460.14) - A(885) = \int_{885}^{460.14} \frac{u}{u - F/E} \cdot \ln\frac{u - F/E}{u} du = 259.36,$$

$$J(885) = \ln(\frac{885 - 211.675}{885}) = -0.27336,$$

and

$$D(460.14) - D(885)) = -635.88.$$

Solving (4), we find that the approximate departure angle is

$$\alpha_0 = \sin^{-1}(0.19603) / 2 = 5.6524° .$$

Improving Accuracy

Time of flight to the target is

$$t = -\frac{1}{\overline{B}g\cos\alpha_0}[T(u) - T(v_0)] = -\frac{1}{0.0064842g\cos(5.6524)}[T(460.143) - T(885)] = 15.75s \text{,}$$

The ordinate of the trajectory vertex is

$$y_{m2} = \frac{g \cdot t^2}{8} = \frac{g \cdot (15.75)^2}{8} = 304m.$$

The value of the average altitude of the trajectory is

$$\bar{y}_2 = 2y_{m2} / 3 = 2 \cdot 304 / 3 = 203m.$$

The value of the density function is

$$h_2(\bar{y}_2) = (\frac{\tau_2 - 0.006328\bar{y}_2}{\tau_2})^{4.4} = (\frac{279.23 - 0.006328 \cdot 203}{279.23})^{4.4} = 0.9954 .$$

Estimate $\overline{B}$ using (3.2.10):

$$\overline{B} = \frac{\bar{c} \cdot E \cdot h(\bar{y}_2)}{g\sqrt{\cos\alpha_0}} = \frac{(0.45002) \cdot (0.1413) \cdot (0.9954)}{g\sqrt{\cos(5.6524)}} = 0.00647 . \quad (5)$$

Comparing (5) and (1), we see that there is a small change in the value of $\overline{B}$. So, the calculated value of the departure angle can be accepted as accurate.

Anyway, we are going to estimate the departure angle for the improved value of the parameter $\overline{B}$.

Using (3.2.5) with the estimated value (4), we have

$$u_T + (211.675) \cdot \ln(u_T - 211.675) = 1629 .$$

Solving the above equation, we find that the corrected value of the pseudo-speed is

$$u_T = 460.90.$$

Substituting in (3.2.12), we obtain

$$\sin(2\alpha_0) = -\frac{2}{(211.675)(0.00647)} \cdot [\frac{A(460.9) - A(885)}{D(460.9) - D(885)} - J(885)] = 0.19576 \cdot$$

Hence, we find that the departure angle is

$$\alpha_0 = \sin^{-1}(0.19576) / 2 = 5.6446° .$$

Note:
The range table of the Russian cannon 122mm, Mod. 1960 shows that the value of the departure angle $\alpha_0 = 5.6333°$, while the maximum altitude of the trajectory is 301 meters.

3.3. Siacci's Trajectories in Non-standard Atmosphere

The following formulas that are derived using the Siacci's solutions of the equations of projectile flight (3.2.5)-(3.2.8) show the relationship between two trajectories of two different projectiles. [64].

We assume that the flight of the projectile is in a non standard atmosphere: the density, the pressure, and the temperature at the firing site are respectively ρ_2, p_2, τ_2.

The density, the pressure and the virtual temperature for the standard atmosphere related to a given G-function are respectively ρ_{0N}, p_{0N}, τ_{0N}.

For a given initial speed of the projectile v_0 and a departure angle α_0 in the interval -15 degrees to 15 degrees, the equations (3.2.5)-(3.2.8), show that the elements of the trajectory at a given point (x, y) depend on the ballistics coefficient $\bar{c}$, and the departure

---

64. Okunev, B. H, Fundamentals of Ballistics, page 223, Vol.1, Book 2, Moscow, 1943

angle α_0 (we assume a constant value of the density function). In other words, all the elements of the trajectory of a projectile fired with an initial speed v_0 are determined by the departure angle α_0 and the ballistics coefficient $\bar{c}$ (or equivalently by the parameter $\bar{B}$, determined in the above section).

Consider two projectiles respectively with ballistics coefficients $\bar{c}_1$ and $\bar{c}_2$ (corresponding parameters $\bar{B}_1$ and $\bar{B}_2$) launched in a non-standard atmosphere with the same initial speed v_0, but with different departure angles, respectively, α_{01} and α_{02}. Consider as well the elements of the first trajectory at a point with coordinates (x_1, y_1) and the elements of the second trajectory at a point with coordinates (x_2, y_2), (figure 8).

We assume that the pseudo-speed u_1 of the projectile at (x_1, y_1) is equal to the pseudo-speed of the projectile at (x_2, y_2), i.e., $u_2 = u_1$.

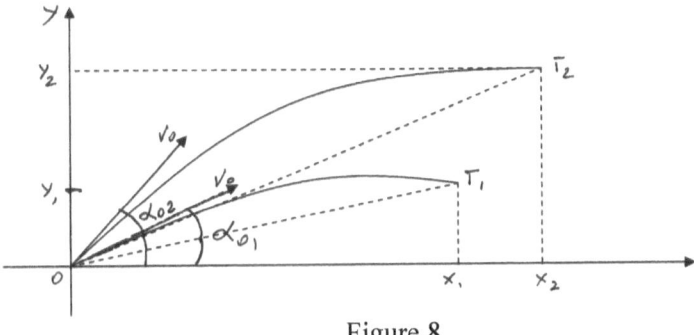

Figure 8

Since the initial speed of both projectiles is v_0, and the pseudo-speed of the first projectile at (x_1, y_1) is equal to the pseudo-speed of the second projectile at (x_2, y_2), employing (3.2.5), (3.2.6), (3.2.7), (3.2.8) for each projectile trajectory, we obtain the following formulae that show the relation of the elements of the second trajectory with the elements of the first trajectory at the corresponding trajectory points:

$$x_2 = \frac{\overline{B}_1}{\overline{B}_2} x_1, \tag{3.3.1}$$

$$y_2 = p_{02} x_2 - \frac{\overline{B}_1^2 \cos^2 \alpha_{01}}{\overline{B}_2^2 \cos^2 \alpha_{02}} p_{01} x_1, \quad \text{since } y_1 = 0,$$

or

$$y_2 = x_2 (p_{02} - \frac{\overline{B}_1 \cos^2 \alpha_{01}}{\overline{B}_2 \cos^2 \alpha_{02}} p_{01}), \tag{3.3.2}$$

$$t_2 = \frac{\overline{B}_1 \cos(\alpha_{01})}{\overline{B}_2 \cos(\alpha_{02})} t_1, \tag{3.3.3}$$

$$p_2 = p_{02} - \frac{\overline{B}_1 \cos^2(\alpha_{01})}{\overline{B}_2 \cos^2(\alpha_{02})} (p_{01} - p_1). \tag{3.3.4}$$

where for the non standard atmosphere,

$$\overline{B} = \frac{\overline{c} \cdot E \cdot h_2(\overline{y}_2)}{g\sqrt{\cos \alpha_0}}, \quad h_2(\overline{y}_2) = (\frac{\tau_2 - 0.006328\overline{y}_2}{\tau_2})^{4.4}, \tag{3.3.5}$$

$$\overline{y}_2 = 2 y_{m2} / 3,$$

or

$$\overline{B} = \frac{\overline{c} \cdot E \cdot h_2(\overline{y}_2)}{g \cdot \cos \alpha_0}, \quad h_2(\overline{y}_2) = (\frac{\tau_2 - 0.006328\overline{y}_2}{\tau_2})^{4.4}, \tag{3.3.6}$$

$$\overline{y}_2 = y_{m2} / 2,$$

respectively, for the field artillery and anti-aircraft shooting (inclined fire).

We can apply the above formulas for two trajectories of the same projectile:

- The first trajectory is at the sea level while the second trajectory is at a given altitude over the sea level.
- Both projectiles are fired at the same site (at the sea level or over it), but the first projectile is fired in a standard atmosphere while the second projectile is fired in a non-standard atmosphere.

The impact point of the projectile fired in a standard atmosphere is (x_{T1}, y_{T1}) while the impact point of the projectile fired in non-standard atmosphere is at (x_{T2}, y_{T2}). Since $y_{T1} = y_{T2} = 0$, i.e., since the impact point of each projectile is on the ground at the same altitude as the muzzle of the respective firearm, then from the equations (3.3.1)-(3.3.4), we obtain:

$$x_{T2} = \frac{\overline{B}_1}{\overline{B}_2} x_{T1}, \tag{3.3.7}$$

$$\sin(2\alpha_{02}) = \frac{\overline{B}_1}{\overline{B}_2} \sin(2\alpha_{01}), \tag{3.3.8}$$

$$t_{T2} = \frac{\overline{B}_1}{\overline{B}_2} \frac{\cos(\alpha_{01})}{\cos(\alpha_{02})} t_{T1}, \tag{3.3.9}$$

$$\sin(2\overline{\alpha}_{T2}) = \frac{\overline{B}_1}{\overline{B}_2} \sin(2\overline{\alpha}_{T1}). \tag{3.3.10}$$

We assume that at the point of impact, the terminal speed of the first projectile and the terminal speed of the second projectile are equal.

Using (3.3.7)-(3.3.10), we can find the elements of the second projectile trajectory when we know the elements of the first trajectory at the point of impact on condition that both projectiles have the same initial speed v_0 and the same terminal speed v_T.

3.4. Siacci's Trajectories and Range Tables

The standard range tables of a firearm usually are constructed for a given standard atmosphere and for firing site at the sea level or close to it. The range tables contain as well the corrections obtained for small deviations of the characteristics of the atmosphere from the standard characteristics at the sea level.

The range tables that are used in the army have as well the corrections for small deviations of the ballistics characteristics of the projectile from the corresponding standard characteristics.

The range tables and the corresponding corrections are not valid when shooting does not take place at the sea level, or when the deviations of the characteristics of the atmosphere from the standard ones are relatively large. For that reason, the army range tables contain as well the range tables for shooting at sites over the sea level, considering that at the sea level the atmosphere is standard.

Applying the relationships between the elements of two trajectories obtained in section 3.3, we can construct the range tables of a projectile at a given altitude when we know the range table of the same projectile at the sea level in standard atmosphere.

Indeed, let's consider the first trajectory as the trajectory of a projectile fired at the sea level in standard atmosphere with initial speed v_0 and departure angle α_{01} and the second trajectory as the trajectory of the same projectile fired in non-standard atmosphere (at the sea level or over the sea level) with the same initial speed v_0 and departure angle α_{02}. The ballistics coefficient of the projectile is

$$c = \frac{id^2}{m} 1000 . \tag{3.4.1}$$

For the field artillery (when the firearm and the target are at the sea level or at the same altitude over the sea level), and for the projectile fired in standard atmosphere, the parameter $\overline{B}$, given by (3.3.5) or (3.3.6), is

$$\overline{B}_1 = \frac{c \cdot E \cdot h_1(\overline{y}_1)}{g\sqrt{\cos \alpha_{01}}}, \tag{3.4.2}$$

where

$$h_1(\overline{y}_1) = (\frac{\tau_{0N} - 0.006328\overline{y}_1}{\tau_{0N}})^{4.4} . \tag{3.4.3}$$

For the projectile fired in a non-standard atmosphere, at the sea level or over the sea level, the parameter $\overline{B}$ is

$$\overline{B}_2 = \frac{\overline{c} \cdot E \cdot h_2(\overline{y}_2)}{g\sqrt{\cos \alpha_{02}}}, \tag{3.4.5}$$

where

$$h_2(\bar{y}_2) = (\frac{\tau_2 - 0.006328\bar{y}_2}{\tau_2})^{4.4}, \quad \bar{c} = \frac{p_2}{p_{0N}}\sqrt{\frac{\tau_{0N}}{\tau_2}} \cdot c. \tag{3.4.6}$$

On condition that $y_{T1} = y_{T2} = 0$, applying (3.3.7)-(3.3.10), for the two trajectories described above, we obtain the following relationships:

$$x_{T2} = j \cdot \frac{h_1(\bar{y}_1)}{h_2(\bar{y}_2)} \cdot \sqrt{\frac{\cos\alpha_{02}}{\cos\alpha_{01}}} \cdot x_{T1}, \tag{3.4.7}$$

$$\sin(2\alpha_{02}) = j \cdot \frac{h_1(\bar{y}_1)}{h_2(\bar{y}_2)} \cdot \sqrt{\frac{\cos\alpha_{02}}{\cos\alpha_{01}}} \cdot \sin(2\alpha_{01}), \tag{3.4.8}$$

$$\sin(2\alpha_{T2}) = j \cdot \frac{h_1(\bar{y}_1)}{h_2(\bar{y}_2)} \cdot \sqrt{\frac{\cos\alpha_{02}}{\cos\alpha_{01}}} \cdot \sin(2\alpha_{T1}), \tag{3.4.9}$$

$$t_{T2} = j \cdot \frac{h_1(\bar{y}_1)}{h_2(\bar{y}_2)} \cdot \sqrt{\frac{\cos\alpha_{02}}{\cos\alpha_{01}}} \cdot t_{T1}, \tag{3.4.10}$$

where

$$j = \frac{p_2}{p_{0N}}\sqrt{\frac{\tau_{0N}}{\tau_2}}. \tag{3.4.11}$$

The trajectory maximum altitude can be found using the approximate formula (3.1.9), i.e.,

$$y_{m2} = \frac{g \cdot t_{T2}^2}{8}.$$

For approximate calculations, we can consider

$$\frac{h_1(\bar{y}_1)}{h_2(\bar{y}_2)} \cdot \sqrt{\frac{\cos\alpha_{02}}{\cos\alpha_{01}}} = 1. \tag{3.4.12}$$

The formulas (3.4.7)-(3.4.11) can be written respectively in the following forms:

$$x_{T2} = j \cdot x_{T1},\tag{3.4.13}$$

$$\sin(2\alpha_{02}) = j \cdot \sin(2\alpha_{01}),\tag{3.4.14}$$

$$\sin(2\alpha_{T2}) = j \cdot \sin(2\alpha_{T1}),\tag{3.4.15}$$

$$t_{T2} = j \cdot t_{T1}.\tag{3.4.16}$$

The above formulae can be used to construct the range table of a projectile in a non standard atmosphere when it is known the range table of the same projectile in standard atmosphere is known.

As it is demonstrated in the following examples, the accuracy of the range tables that can be obtained using the above method is acceptable.

Note the similarity of the formulas obtained in this section with the formulas obtained using the Snell's law and the similitude trajectories (chapter 2).

Summary

The example 4.1 (below) shows that for the Siacci's trajectories, we can use the following formulae:

$$x_{T2} = j \cdot x_{T1},\tag{3.4.17}$$

$$\alpha_{02} = \frac{1}{2} ar \sin(j \cdot \sin(2\alpha_{01})),\tag{3.4.18}$$

$$v_2 = v_1,\tag{3.4.19}$$

$$t_2 = j \cdot t_1,\tag{3.4.20}$$

$$x_{m2} = j \cdot x_{m1}, \quad y_m = \frac{g \cdot t_T^2}{8},\tag{3.4.21}$$

where

$$j = \frac{p_{0N}}{p_0} \sqrt{\frac{\tau_0}{\tau_{0N}}} \qquad (3.4.22)$$

Note that the expression

$$\frac{h_1(\bar{y})}{h_2(\bar{y})} \cdot \sqrt{\frac{\cos \alpha_{02}}{\cos \alpha_{01}}}$$

on the right side of formulae (3.4.7)-(3.4.10) represents the constant quotient β_1 / β_2, i.e.,

$$\frac{\beta_1}{\beta_2} = \frac{h_1(\bar{y}_1)}{\sqrt{\cos \alpha_{01}}} \div \frac{h_2(\bar{y}_2)}{\sqrt{\cos \alpha_{02}}} . \qquad (3.4.23)$$

Example 4.1 High-Altitude shooting

Use the table of the Russian 122mm cannon Mod.1960 to find the departure angle needed to hit the target located at a range of 10,000 meters if the firing site is 1,500 meters over the sea level.

The temperature, the virtual temperature, and the pressure at the firing site (1,500 meters over the sea level) are respectively: 6 degrees Celsius, 279.748 degrees Kelvin, and 626 mmHg. [65]

The initial speed of the projectile is 885m/s.

The range table of the 122mm cannon is obtained for the TSA atmosphere: The temperature, the virtual temperature and the pressure at the sea level (firing site) are respectively 15 degrees Celsius, 289.08 degrees Kelvin, and 750 mmHg.

Solution:

As a first approach, we consider that

$$\frac{h_1(\bar{y}_1)}{h_2(\bar{y}_2)} \cdot \sqrt{\frac{\cos \alpha_{02}}{\cos \alpha_{01}}} = 1 .$$

---

65. Ministry of Defense of Albania, "*Range Tables of 122mm cannon Model 1960*", Tirana, 1967

The value of j is:

$$j = \frac{p_{0N}}{p_2}\sqrt{\frac{\tau_2}{\tau_{0N}}} = \frac{750}{626}\sqrt{\frac{279.748}{289.08}} = 1.17859.$$

The range of the projectile in standard atmosphere that has the same terminal pseudo-speed as the projectile fired at 1,500 meters over the sea level, is approximately

$$x_{T1} = j^{-1} \cdot x_{T2} = 0.84847 \cdot (10{,}000) = 8485m.$$

From the range table of the 122mm cannon (sea level range tables, standard TSA atmosphere), [8], by interpolation, we find that the departure angle needed to hit the target located at the horizontal range of 8485 meters is $\alpha_{01} = 4.7875°$.

The Departure Angle

Substituting in (3.4.14), we have:

$$\sin(2\alpha_{02}) = 1.17859 \cdot \sin(2 \cdot 4.7875).$$

Solving the above equation, we find that the departure angle needed to hit the target at the range of 10,000 meters (altitude of the firing site of 1,500 meters over the sea level), is

$$\alpha_{02} = 5.6529°.$$

The Time of Flight

Time of flight to the target located at the sea level is 13.5 seconds. Using (3.4.10), we find that the time of flight to the distance of 10,000 meters is

$$t_{T2} = J \cdot t_{T1} = 1.17859 \cdot (13.5) = 15.91 \ s.$$

Improving accuracy
Maximum Altitude of the Trajectory

The maximum height of the trajectory at the sea level and the maximum height of the trajectory at the altitude of 1,500 meters over the sea level, is respectively approximately to

$$y_{T1} = g\frac{t_1^2}{8} = g\frac{13.5^2}{8} = 223m ., \qquad y_2 = g\frac{t_2^2}{8} = g\frac{(15.91)^2}{8} = 310.29m .$$

Using the above values, we have:

$$\bar{y}_1 = 2y_1/3 = 2 \cdot 203/3 = 149m ,$$

$$h_1(\bar{y}_1) = (\frac{\tau_1 - 0.006328\bar{y}_1}{\tau_1})^{4.4} = (\frac{289.08 - 0.006328 \cdot 149}{289.08})^{4.4} = 0.9805 ,$$

$$\bar{y}_2 = 2y_2/3 = 2 \cdot 310/3 = 207m ,$$

$$h_2(\bar{y}_2) = (\frac{\tau_2 - 0.006328\bar{y}_2}{\tau_2})^{4.4} = (\frac{279.748 - 0.006328 \cdot 207}{279.748})^{4.4} = 0.9796 .$$

We find as well:

$$j \cdot \frac{h_1(\bar{y}_1)}{h_2(\bar{y}_2)} \cdot \sqrt{\frac{\cos\alpha_{02}}{\cos\alpha_{01}}} = (1.17859)\frac{0.9805}{0.9796}\sqrt{\frac{\cos(5.6529)}{\cos(4.79271)}} = 1.17886 .$$

The range of the projectile in standard atmosphere that has the same terminal pseudo-speed as the projectile fired 1,500 meters over the sea level, is

$$x_{T1} = j^{-1} \cdot x_{T2} = 0.8483 \cdot (10,000) = 8483m .$$

The obtained range is quite the same as the approximate range we have obtained above (8,485m). Thus, the approximate results that we have already obtained represent the solution of the problem.

Note that the departure angle (that corresponds to the given range of 10,000 meters , altitude of 1,500 meters) given in the mountain range table of the 122mm Russian cannon Model 1960, is

$$\alpha_{02} = 5.6333°.$$

while the departure angle we have obtained using the Siacci's method is

$$\alpha_{02} = 5.6529°.$$

The difference between those values is

$$\Delta\alpha_{02} = (5.6529° - 0.6333°) = 0.196° = 1.18'.$$

The above example shows that for the Siacci's trajectories, we can use the following formulae:

$$\alpha_{02} = \frac{1}{2}ar\sin(j \cdot \sin(2\alpha_{01})), \tag{1}$$

$$v_2 = v_1, \tag{2}$$

$$t_2 = j \cdot t_1, \tag{3}$$

$$y_m = \frac{g \cdot t_T^2}{8}, \tag{4}$$

where

$$j = \frac{p_{0N}}{p_0}\sqrt{\frac{\tau_0}{\tau_{0N}}}. \tag{5}$$

3.5. The Siacci's Method for Low Speeds

When the projectile is launched with an initial speed of less than 256m/s, the drag function (3.1.2) is

$$G_D(v) = A \cdot v^2, \tag{3.5.1}$$

where the value of the parameter A, is given in sections (1.4), (1.5), and (1.6) for any of the G-functions of resistance.

Using the Siacci's pseudo-speed,

$$u = v\frac{\cos\alpha}{\cos\alpha_0}, \tag{3.5.2}$$

and following the same procedure as in section (5.3) of the "*Exterior Ballistics with Applications*", for a projectile launched at an angle not greater than 15 degrees, the differential equations (3.1.1) that describe the projectile flight can be written:

$$\begin{cases} \dfrac{dp}{du} = \dfrac{1}{Bu^3 \cdot \cos^2\alpha_0} \\[2mm] \dfrac{dt}{du} = -\dfrac{1}{Bgu^2 \cos\alpha_0} \\[2mm] \dfrac{dx}{du} = -\dfrac{1}{Bgu} \\[2mm] \dfrac{dy}{du} = -p\dfrac{1}{Bgu} \end{cases}, \tag{3.5.3}$$

where:

$$B = A \cdot \beta\frac{c}{g}, \qquad c = \frac{id^2}{m}1000, \qquad p = \tan\alpha, \tag{3.5.4}$$

and

(a)   For anti-aircraft (or downhill/uphill) fire:

$$\beta = h(\bar{y}) / \cos\alpha_0. \tag{3.5.5}$$

(b)   For field artillery fire:

$$\beta = \frac{h(\bar{y})}{\sqrt{\cos\alpha_0}}. \tag{3.5.6}$$

In a similar way as in section 5.3 of the "*Exterior Ballistics with Applications*", from the system of differential equations (3.5.3), we obtain the following analytical formula that describe the flight of the projectile for speeds of less than 256m/s in a standard atmosphere (ASM, ICAO, TSA):

- The equation of the projectile trajectory:

$$y = \tan\alpha_0 \cdot x + \frac{x}{2B\cos^2\alpha_0}\left[\frac{1}{2}\left(\frac{1}{u^2} - \frac{1}{v_0^2}\right)\cdot\left(\frac{1}{\ln u - \ln v_0}\right) + \frac{1}{v_0^2}\right].\qquad (3.5.7)$$

- The abscissa of the projectile:

$$x = -\frac{1}{Bg}(\ln u - \ln v_0).\qquad (3.5.8)$$

- The angle of flight:

$$\tan\alpha = \tan\alpha_0 - \frac{1}{2B\cos^2\alpha_0}\left(\frac{1}{u^2} - \frac{1}{v_0^2}\right).\qquad (3.5.9)$$

- The time of flight:

$$t = \frac{1}{Bg\cos\alpha_0}\left(\frac{1}{u} - \frac{1}{v_0}\right).\qquad (3.5.10)$$

Notes:

- As we have pointed out in the comment in page 264 of "*Exterior Ballistics with Applications*" (i.e. in Example 1, page 260-264), the Siacci's method can be used as well for the projectile launched with any angle if the initial speed of the projectile is not greater than 256 m/s.
- The above formulae can be further simplified considering the notes at the aend of section 3.1.

Example 5.1 The Horizontal Range of a 105mm Mortar Projectile

A projectile with a of mass 14.97 kg is fired at the sea level from a 105mm mortar with a speed of 205m/s (charge 1) at angle of 45° 45 degrees. Consider the ICAO atmosphere.

Determine the point of impact of the projectile on the ground at the sea level, time of flight, the velocity of impact, the terminal angle, and the maximum height of the flight.

The ballistics coefficient of projectile relative to the G_1 -function of resistance (ICAO atmosphere) is 0.4269. The atmospheric conditions are normal.

Solution:

The drag function displayed in the formula (1.5.4) is

$$G_1(v) = 1.0584 \times 10^{-4} \cdot v^2.$$

The value of A is

$$A = 1.0584 \times 10^{-4}$$

Considering as a first approach $h(\bar{y}) \approx 1$, we find:

$$\beta = \frac{h(\bar{y})}{\sqrt{\cos \alpha_0}} = \frac{1}{\sqrt{\cos 45°}} = 1.18921$$

and

$$B = A \cdot \beta \frac{c}{g} = 1.0584 \cdot 10^{-4}(1.18921)(0.4269)/(9.80665) = 5.4792 \cdot 10^{-6}$$

The Maximum Height of the Projectile

Maximum height of the projectile is located at the point where $\tan \alpha = \tan(0°) = 0$. Substituting the above value in (3.5.9), we have:

$$p_0 - \frac{1}{2B \cos^2 \alpha_0}(\frac{1}{u^2} - \frac{1}{v_0^2}) = 0.$$

Substituting in the above equation the required values, we can write:

$$1 - \frac{1}{2(5.4792 \cdot 10^{-6}) \cos^2(45°)} \left(\frac{1}{u^2} - \frac{1}{205^2}\right) = 0$$

Hence, we find that the pseudo-speed at the vertex of trajectory is $u_m = 184.82 m/s$. Substituting in (3.5.8), we find the abscissa of the trajectory vertex:

$$x_m = -\frac{1}{Bg}(\ln u_m - \ln v_0) =$$

$$= -\frac{1}{(5.4792 \cdot 10^{-6})(9.80665)}[\ln(184.82) - \ln(205)] = 1928.33 m$$

Substituting in the equation of the trajectory (3.5.7) we find that the y-coordinate of the vertex is

$$y_m = \tan\alpha_0 \cdot x + \frac{x}{2B\cos^2\alpha_0}\left[\frac{1}{2}\left(\frac{1}{u_m^2} - \frac{1}{v_0^2}\right) \cdot \left(\frac{1}{\ln u_m - \ln v_0}\right) + \frac{1}{v_0^2}\right] = 1928.33 \cdot \tan(45) +$$

$$+ \frac{1928.33}{2(5.4792 \cdot 10^{-6})\cos(45°)^2}\left[\frac{1}{2}\left(\frac{1}{184.82^2} - \frac{1}{205^2}\right) \cdot \frac{1}{\ln(184.82) - \ln(205)} + \frac{1}{205^2}\right] = 997 m$$

The Value of B.

We find:

$$\bar{y} = (2/3)y_m = (2/3)(997) = 665$$

and

$$h(\bar{y}) = \left(\frac{288.15 - 0.006328\bar{y}}{288.15}\right)^{4.4} = \left(\frac{288.15 - 0.006328 \cdot (665)}{288.15}\right)^{4.4} = 0.9373$$

The value of β is

$$\beta = \frac{h(\bar{y})}{\sqrt{\cos\alpha_0}} = \frac{0.9373}{\sqrt{\cos(45°)}} = 1.11467,$$

while

$$B = A \cdot \beta \frac{c}{g} = 5.13571 \times 10^{-6}.$$

The Range and the Time of Flight

Method 1

The range x is at the point where the y-coordinate is zero. Substituting $y=0$, in the equation of trajectory (3.5.7), we have:

$$\tan \alpha_0 + \frac{1}{2B\cos^2 \alpha_0}[\frac{1}{2}(\frac{1}{u^2} - \frac{1}{v_0^2}) \cdot (\frac{1}{\ln u - \ln v_0}) + \frac{1}{v_0^2}] = 0.$$

Substituting the required values, we get the following equation:

$$\tan(45) + \frac{1}{2(5.1357 \cdot 10^{-6})\cos^2(45°)}[\frac{1}{2}(\frac{1}{u^2} - \frac{1}{(205)^2}) \cdot (\frac{1}{\ln u - \ln(205)}) + \frac{1}{(205)^2}] = 0 \cdot$$

Solving the last equation for the pseudo-speed u using a TI-83Plus graphing calculator we find

$$u = 169.62 m / s.$$

Substituting the value of the pseudo-speed in equations (3.5.8) and (3.5.10), we find that the range and the time flight to the target are respectively:

$$x_T = -\frac{1}{Bg}(\ln u - \ln v_0) = -\frac{1}{(5.1357 \cdot 10^{-6})(9.80665)}(\ln 169.62 - \ln 205) = 3761.60m$$

and

$$t_T = \frac{1}{(5.1357 \cdot 10^{-6})(9.80665)\cos(45°)}(\frac{1}{169.62} - \frac{1}{205}) = 28.52s.$$

Method 2

We can avoid the procedure shown in method 1 and the equation (3.5.7), using the relation (3.1.9) to find the time of flight, i.e., employing

$$y_m = \frac{g \cdot t_T^2}{8}.$$

We find that the time of flight to the target is

$$t_T = (8 \cdot y_m / g)^{1/2} = [8 \cdot (997) / 9.80665]^{1/2} = 28.52s.$$

Employing the equation (3.5.10),

$$t = \frac{1}{Bg \cos \alpha_0} (\frac{1}{u} - \frac{1}{v_0}),$$

we find that the pseudo-speed at the impact point on the ground is:

$$u_T = \frac{v_0}{1 + Bgtv_0 \cos \alpha_0} = \frac{205}{1 + (5.1357 \cdot 10^{-6}) \cdot (g)(12.52) \cdot (205) \cos(45°)} = 169.67 m/s$$

The Terminal Angle

Substituting in (3.5.9), we find:

$$\tan \alpha = \tan \alpha_0 - \frac{1}{2B \cos^2 \alpha_0} (\frac{1}{u^2} - \frac{1}{v_0^2}) = \tan(45°) -$$

$$- \frac{1}{2 \cdot (5.13571 \cdot 10^{-6}) \cos^2 (45°)} \cdot (\frac{1}{169.67^2} - \frac{1}{205^2}) = -1.13046.$$

Hence, for the impact angle, we find:

$$\alpha_T = \tan^{-1}(-1 1.13046) = -48.38847657° = -48.5043°.$$

The Terminal Speed

Employing the relation (3.5.2), we find that the terminal speed of the projectile, at the point of impact to the target, is

$$v = u \frac{\cos \alpha_0}{\cos \alpha} = 169.62 \frac{\cos(45°)}{\cos(-48.50427°)} = 181.12 m/s.$$

In the following table are given, for comparison, the data obtained using the PC program Rprojc.Bas.

Table 1

	Launching Angle	Angle of impact	Range (meter)	Time of flight(s)	Height (meter)
Siacci method	$\alpha_0 = 45°$	$\alpha_T = -48.50°$	$x_T = 3756$	$t_T = 28.52$	$y_m = 997$
Rprojc.Bas	$\alpha_0 = 45°$	$\alpha_T = -48.35°$	$x_T = 3760$	$t_T = 28.48$	$y_m = 995$

Example 5.2 Departure Angle, 120mm Mortar Projectile

A projectile of with a mass of 13.59 kg is fired from a 120mm cannon with a speed of 102m/s, at the sea level in the ICAO atmosphere. Find the launching angle needed to hit a target located at a horizontal distance of 770m.

The ballistics coefficient with respect to G_2-function of resistance is 0.6978.

Determine as well all the other elements of trajectory at the point of impact and the maximum height of the flight trajectory.

Solution:

The G_2-function of resistance for the projectile fired with the initial speed of less than 256m/s in the ICAO atmosphere, displayed in formula (1.5.6), is

$$G_1(v) = 9.2868 \times 10^{-5} \cdot v^2 .$$

The value of A is

$$A = 9.2868 \times 10^{-5} .$$

I. **Approximate Solution**

Since the maximum height of the trajectory and the launching angle are unknown, as a first approach, we consider $h(\bar{y})=1$, and $\cos\alpha_0 =1$. Thus,

$$\beta = h(\bar{y}) / \sqrt{\cos\alpha_0} = 1$$

and

$$B = A \cdot \beta \frac{c}{g} = 9.2868 \cdot 10^{-5} (0.6978) / (9.80665) = 6.6080965 \cdot 10^{-6}.$$

The Pseudo-Speed at the Point of Impact

Solving for $\ln u$ the relation (3.5.8),

$$x = -\frac{1}{Bg}(\ln u - \ln v_0),$$

we have:

$$\ln u = \ln v_0 - B \cdot g \cdot x = \ln(102) - (6.608966 \cdot 10^{-6}) \cdot (9.80665) \cdot (770) = 4.57507.$$

Hence, for the pseudo-speed, we find:

$$u = e^{4.57507} = 97.035 m / s.$$

The Departure Angle

At the point of impact, the y-coordinate is zero. Substituting $y = 0$ on the right side of (3.5.7) as well as all required values, and considering that

$$\cos^2 \alpha_0 = (1 + \tan^2 \alpha_0)^{-1} = (1 + p_0^2)^{-1},$$

we have:

$$\tan(\alpha_0) + \frac{(1 + \tan^2 \alpha_0)}{2(6.6080965 \cdot 10^{-6})}[\frac{1}{2}(\frac{1}{97^2} - \frac{1}{102^2}) \cdot \frac{1}{\ln(97) - \ln(102)} + \frac{1}{102^2}] = 0.$$

Hence, we can write:

$$\tan \alpha_0 - 0.3781(1 + \tan \alpha_0^2) = 0.$$

Hence, we find two values for $\tan \alpha_0$:

$$\tan \alpha_0 = 2.1877, \qquad \tan \alpha_0 = 0.4571.$$

For the corresponding launching angle, we obtain respectively

$$\alpha_0 = \tan^{-1}(2.1877) = 65.4349°,$$

and

$$\alpha_0 = \tan^{-1}(0.4571) = 24.5651°.$$

Let's solve the problem for the biggest launching angle $\alpha_0 = 65.4349°$. In a similar way, we can solve the problem for the other angle.

II. Improving Results

The Coordinates of Trajectory Vertex

Using the first relation of (3.1.8), we find that the x-coordinate of the trajectory vertex is

$$x_m = (0.5 + 0.0001 \cdot v_0) \cdot x_T = (0.5 + 0.0001 \cdot 102) \cdot 770 = 392.85m.$$

Solving (3.5.8),

$$x = -\frac{1}{Bg}(\ln u - \ln v_0),$$

for "$\ln u$" we have:

$$\ln u_m = \ln v_0 - B \cdot g \cdot x_m = \ln(102) - (6.608966 \cdot 10^{-6}) \cdot (9.80665) \cdot (392.85) = 4.5995.$$

Hence, for the pseudo-speed, at the trajectory vertex we find:

$$u_m = e^{4.5995} = 99.44m/s.$$

Substituting in the equation of the trajectory (3.5.7), we find that the y-coordinate of the vertex is

$$y_m = \tan\alpha_0 \cdot x + \frac{x}{2B\cos^2\alpha_0}[\frac{1}{2}(\frac{1}{u_m^2} - \frac{1}{v_0^2}) \cdot (\frac{1}{\ln u_m - \ln v_0}) + \frac{1}{v_0^2}] = 392.85 \cdot \tan(65.4349) +$$

$$+ \frac{392.85}{2(6.6081 \cdot 10^{-6})\cos(65.435°)^2}[\frac{1}{2}(\frac{1}{99.44^2} - \frac{1}{102^2}) \cdot \frac{1}{\ln(99.44) - \ln(102)} + \frac{1}{102^2}] = 432m$$

We find that

$$\bar{y} = (2/3)y_m = (2/3)(432) = 288m$$

and

$$h(\bar{y}) = (\frac{288.15 - 0.006328\bar{y}}{288.15})^{4.4} = (\frac{288.15 - 0.006328 \cdot (288)}{288.15})^{4.4} = 0.9725.$$

The new value of β is

$$\beta = h(\bar{y}) / \sqrt{\cos\alpha_0} = 0.9725 / \sqrt{\cos(65.4349°)} = 1.50824$$

and

$$B = A \cdot \beta\frac{c}{g} = 9.2868 \cdot 10^{-5} \cdot (1.50824) \cdot (0.6978) / (9.80665) = 9.9666 \cdot 10^{-6}$$

The Pseudo-Speed at the Point of Impact, Horizontal Range 770m

Solving for $\ln u$ the equation (3.5.8),

$$x = -\frac{1}{Bg}(\ln u - \ln v_0),$$

we have:

$$\ln u = \ln v_0 - B \cdot g \cdot x = \ln(102) - (9.9666 \cdot 10^{-6}) \cdot (9.80665) \cdot (770) = 4.5497.$$

Hence, for the pseudo-speed, we find:

$$u = e^{4.5497} = 94.60m/s.$$

The Departure Angle

At the point of impact, the y-coordinate is zero. Substituting $y = 0$ on the right side of (3.5.7) as well as all required values, and considering that

$$\cos^2\alpha_0 = (1 + \tan^2\alpha_0)^{-1} = (1 + p_0^2)^{-1},$$

we have:

$$\tan(\alpha_0) + \frac{(1 + \tan^2 \alpha_0)}{2(9.9666 \cdot 10^{-6})} [\frac{1}{2}(\frac{1}{94.60^2} - \frac{1}{102^2}) \cdot \frac{1}{\ln(94.60) - \ln(102)} + \frac{1}{102^2}] = 0$$

Hence, we can write:

$$\tan \alpha_0 - 0.38211(1 + \tan \alpha_0^2) = 0.$$

Hence for $\tan \alpha_0$, we find:

$$\tan \alpha_0 = 2.15247.$$

The corresponding launching angle is

$$\alpha_0 = \tan^{-1}(2.15247) = 65.08°.$$

The time of flight to the target is

$$t_T = \frac{1}{Bg \cos \alpha_0}(\frac{1}{u} - \frac{1}{v_0}) = \frac{1}{(9.967 \cdot 10^{-6})(9.8066) \cos(65.08°)}(\frac{1}{94.60} - \frac{1}{102}) = 18.62s$$

The y-coordinate of the trajectory vertex is

$$y_m = \frac{g \cdot t_T^2}{8} = \frac{9.80665 \cdot (18.62)^2}{8} = 425m.$$

The corresponding x-coordinate found above is

$$x_m = 392.85m$$

Estimation of the Other Elements of the Trajectory

Using the data obtained so far, hereafter we estimate the other elements of the trajectory.

The Impact Angle

Employing (3.5.9), we have:

$$\tan \alpha_T = \tan \alpha_0 - \frac{1}{2B\cos^2 \alpha_0}(\frac{1}{u^2} - \frac{1}{v_0^2}) =$$

$$= \tan(60.08°) - \frac{1}{2(9.9666 \cdot 10^{-6})\cos^2(60.08°)}(\frac{1}{94.60^2} - \frac{1}{102^2}) = -2.263$$

Hence, we find that the impact angle is

$$\alpha_T = \tan^{-1}(-2.263) = -66.1598°$$

The Impact Speed

Employing the pseudo-speed formula, i.e., the third formula of (4), we can find

$$v_T = u_T \frac{\cos \alpha_0}{\cos \alpha_T} = 94.60 \frac{\cos(65.08°)}{\cos(-66.1598°)} = -98.62 m/s.$$

In the following table are given for comparison the data obtained above, with the data obtained employing the PC program Rprojc.Bas.

Table 2: Mortar 120mm

Method	Launching Angle	Angle of impact	Impact speed	Time of flight	Maximum Height
Siacci	65.08	-66.16	98.62	18.62	432
Aprojc.Bas	65.00	-65.96	98	18.55	422

3.6. Motion of Projectiles Launched at Small Departure Angles

For small departure angles, the solution of the differential equations of projectile flight can be obtained in analytical form. For such angles, the trajectory of the projectile flight can be described with an equation.

Indeed, we consider the system of differential equations of projectile flight, (2.1.3), written in a similar way as (2.4.12), i.e., the system:

$$\begin{cases} \dfrac{dv_x}{dt} = -\bar{c} \cdot h_0(y) \cdot G_D(v) \cdot \dfrac{v_x}{v} \\ \dfrac{dp}{dt} = -\dfrac{g}{v_x} \\ \dfrac{dx}{dt} = v_x \\ \dfrac{dy}{dt} = v_y \end{cases} \quad , \quad p = \tan\alpha \qquad (3.6.1)$$

where

$$G_D(v) = \begin{cases} A \cdot v^2 & for \quad v \le 256m/s \\ E \cdot (v - F/E) & for \quad v > 256m/s \end{cases}, \qquad (3.6.2)$$

$$h_0(y) = (\frac{\tau_0 - 0.006328y}{\tau_0})^{4.4}, \qquad\qquad (3.6.3)$$

$$\bar{c} = \frac{p_0}{p_{0N}} \sqrt{\frac{\tau_{0N}}{\tau_0}} \cdot c, \qquad c = \frac{id^2}{m} 1000. \qquad (3.6.4)$$

General Approximations

To obtain simple analytical formulas, we integrate the system of differential equations of projectile flight, making a series of approximations that affect slightly the accuracy of solutions.

Thus, we assume that the density function $h_0(y)$ is a constant and is determined by the relation

$$h_0(y) = h_0(\bar{y}), \qquad\qquad (3.6.5)$$

where

- $\bar{y} = (2/3)y_m$, for field artillery,
- $\bar{y} = (1/2)y_m$, for anti-aircraft artillery (up-hill, or down-hill shooting), y_m is the maximum height of the trajectory.

We can write:

$$h_0(\bar{y}) = (\frac{\tau_0 - 0.006328\bar{y}}{\tau_0})^{4.4} \tag{3.6.6}$$

In case the ballistics temperature, ballistics density, and ballistics pressure are known (given by a metro-station bulletin) we substitute in (3.6.3)

$$h_0(y) = \rho_B / \rho_0, \tag{3.6.7}$$

where ρ_B and ρ_0 are respectively the ballistics density and the density at the departing point of projectile.

We see that the approximations (3.6.6) and, (3.6.7) eliminate the dependence of the differential equations from the density function.

Approximations in Velocity

When a projectile is launched with a relatively small angle (for example in case of shooting with assaulting riffles, automatic guns, etc., antitank -firearmsing, or in case of flight of fragments of antipersonnel ammunitions or of antipersonnel mines, etc.) we can consider that the projectile speed "v" is approximately equal to the x-component of the velocity, v_x, i.e.,

$$v = (v_x^2 + v_y^2)^{1/2} = v_x(1 + v_y^2/v_x^2)^{1/2} \approx v_x. \tag{3.6.8}$$

The above condition is satisfied if at any time along the projectile trajectory, we have:

$$v_y^2 / v_x^2 = p^2 \ll 1, \tag{3.6.9}$$

or

$$\tan^2 \alpha \ll 1. \tag{3.6.10}$$

The projectile position during the flight is relatively close to the ground (site) where the projectile is launched. As result, we can assume that the density function $h_0(y)$ is constant and equal to the value it has at the departure point. In standard atmosphere, at the sea level we have $h_0(y) = h_0(\bar{y}) = 1$.

Note that for the flight of projectile close to the ground, the density function have practically the value it has at the muzzle of the firearm.

Thus, considering the following approximations,:

- the projectile speed v at any time from launching point till it hits the target is greater than 256m/s, $v > 256m/s$,.
- during the projectile flight, at any time, $p^2 = \tan^2 \alpha << 1$, and $v \approx v_x$,.
- The density function $h_0(y)$ is constant and equal to the value it has at the launching point,

for high-speed projectiles ($v > 256m/s$) the system of equations (3.6.1) can be written:

$$\begin{cases} \dfrac{dv_x}{dt} = -b \cdot (v_x - F/E) \\ \dfrac{dp}{dt} = -\dfrac{g}{v_x} \\ \dfrac{dx}{dt} = v_x \\ \dfrac{dy}{dt} = v_y \end{cases} \qquad , \qquad (3.6.11)$$

where

$$b = \bar{c} \cdot E \cdot h_0(\bar{y}), \qquad \bar{c} = \dfrac{p_0}{p_{0N}} \sqrt{\dfrac{\tau_{0N}}{\tau_0}} \cdot c, \qquad c = i\dfrac{d^2}{m} \cdot 1000, \qquad p = \tan\alpha \cdot$$

$$(3.6.12)$$

Integrating the system of differential equations (3.6.11), we obtain:

The Parametric Equations of Trajectory

The coordinates of the projectile in flight, as a function of time are:

$$x = (F/E) \cdot t + \frac{(v_{x0} - F/E)}{b} \cdot (1 - e^{-b \cdot t}), \qquad (3.6.13)$$

and

$$y = p_0 [(F/E)t + \frac{v_{x0} - F/E}{b \cdot g} \cdot (1 - e^{-b \cdot t})] -$$

$$- \frac{g}{(F/E) \cdot b} \int_0^t (F/E + (v_{x0} - F/E) \cdot e^{-b \cdot t}) \cdot \ln(\frac{(F/E)e^{b \cdot t} + v_{x0} - F/E}{v_{x0}}) dt \qquad (3.6.14)$$

Employing (3.6.13), the equation (3.6.14) can be written in the form

$$y = \tan \alpha_0 \cdot x -$$

$$- \frac{g}{(F/E) \cdot b} \int_0^t (F/E + (v_{x0} - F/E) \cdot e^{-b \cdot t}) \cdot \ln(\frac{(F/E)e^{b \cdot t} + v_{x0} - F/E}{v_{x0}}) dt$$

The Projectile Speed

The components of the velocity are:

$$v_x = F/E + (v_{0x} - F/E) \cdot e^{-b \cdot t}, \qquad (3.6.15)$$

and

$$v_y = v_x \cdot \tan \alpha . \qquad (3.6.16)$$

The speed of the projectile is

$$v = (v_x^2 + v_y^2)^{1/2} \approx v_x . \qquad (3.6.17)$$

Note that the speed of projectile at any time during the flight is approximately equal to the x-component, v_x, of the velocity.

The Projectile Angle

The angle that the projectile forms with the x-axis at time t is

$$\tan \alpha = \tan \alpha_0 - \frac{g}{(F/E) \cdot b} \cdot \ln(\frac{(F/E) \cdot e^{b \cdot t} + v_{0x} - F/E}{v_{0x}}). \quad (3.6.18)$$

Using the above formulae, we are able to find the elements of the projectile trajectory at any moment during the flight.

Example 6.1 Estimation of the Departure Angle

A NATO 5.56×45mm M855 ball bullet is fired from an M16 rifle. The initial speed of the bullet is 945m/s. Find the departure angle needed to hit a target located at the horizontal range of 800 meters. The firearm and the target are at the sea level. The ballistics coefficient relative to the G_1-function is $c = 5.0051m^2 / kg$ ($C = 0.2842lb / in^2$).

Find as well the time of flight, terminal speed, and the terminal angle. Assume that the projectile flight is in the ASM atmosphere, at the sea level.

(Note that this example is identical to example 1.3, section 3.1.)

Solution:

The G_1-function for the ASM atmosphere is

$$G_1(v) = \begin{cases} 1.00347 \times 10^{-4} \cdot v^2 & for \quad v \le 256m/s \\ 0.312914v - 79.3976 & for \quad 256 < v \le 1000 \end{cases}$$

We have:

$$E = 0.312914, \quad F = 79.3976, \text{ and } \quad F/E = 253.74,$$

$$\bar{c} = c, \quad b = E \cdot c = (0.312914) \cdot (5.0051) = 1.5661,$$

and

$$v_{x0} = v_0 \cos \alpha_0 = v_0 \approx 945m/s$$

The Time of Flight to the Target

Substituting in (5.13), we write:

$$800 = (253.74)t + \frac{(945 - 253.74)}{1.5661}(1 - e^{-1.5661 \cdot t})$$

The above equation can be written:

$$800 - (253.74)t = 441.389 \cdot (1 - e^{-1.5661 \cdot t})$$

Using a graphing calculator, we find the time of flight

$$t = 1.564s.$$

The Departure Angle

Substituting $y = 0$ in (5.14), we have:

$$32417.37 \cdot p_0 - \int_0^{1.56}(253.7 + (691.256) \cdot e^{-1.5661 \cdot t}) \cdot \ln(\frac{(253.7)e^{1.5661 \cdot t} + 691.26}{945})dt = 0 \cdot$$

The value of the integral on the right side, calculated using a graphing calculator TI83+, is 348.71. Substituting in the above equation the obtained value of the integral, we obtain the equation:

$$32417.37 \cdot \tan\alpha_0 - 348.71 = 0.$$

Hence we find that the departure angle is

$$\alpha_0 = \tan^{-1}(0.010757) = 0.6163°.$$

The Terminal Speed

$$v_T \approx v_x = F/E + (v_{0x} - F/E) \cdot e^{-b \cdot t} =$$
$$= 253.74 + (945 - 253.74) \cdot e^{-1.5661(1.564)} = 313.42m/s.$$

The Terminal Angle

Substituting in (5.18), we have

$$\tan \alpha_T = \tan \alpha_0 - \frac{g}{(F/E) \cdot b} \cdot \ln\left(\frac{F/E \cdot e^{b \cdot t} + v_{0x} - F/E}{v_{0x}}\right) =$$

$$= \tan(0.6163) - \frac{9.80665}{253.7 \cdot (1.5661)} \cdot \ln\left(\frac{253.7 \cdot e^{1.5661 \cdot (1.564)} + 945 - 253.7}{945}\right) .$$

Hence, for the terminal angle we obtain

$$\alpha_T = \tan^{-1}(-0.022454) = -1.2863° .$$

Note that comparing the results obtained above with the results obtained in example 1.3, section 3.1, using the Siacci's method, we see that they are almost identical.

3.7. Exterior Ballistics of Low-Flying Projectiles

The trajectories of projectiles fired from small arms within their effective range, or the trajectories of artillery projectiles used for direct close fire against some targets (tanks, vehicles, ships, etc.) are close to the ground. For such trajectories, the change in y-coordinate of a projectile during the flight is small for relatively large horizontal distances.

To find the solution of the differential equations of flight assuming low flights, we can approximate the "y-coordinate" of the projectile using the Maclauren series.

Indeed, considering the assumptions made in the preceding section, the system of differential equations (2.4.12), for projectile speeds $v > 256m/s$, can be written in the following form:

$$\begin{cases} \dfrac{dv_x}{dx} = -b\dfrac{v_x - F/E}{v_x} \\ \dfrac{dp}{dx} = -\dfrac{g}{v_x^2} \\ \dfrac{dt}{dx} = \dfrac{1}{v_X} \\ \dfrac{dy}{dx} = p \end{cases} , \qquad v \approx v_x \qquad (3.7.1)$$

where

$$b = E \cdot \overline{c} \cdot h_0(\overline{y}), \quad \overline{c} = \frac{p_0}{p_{0N}} \sqrt{\frac{\tau_{0N}}{\tau_0}} , \quad c = i \frac{\cdot d^2}{m} \cdot 1000 , \qquad (3.7.2)$$

$$h_0(\overline{y}) = (\frac{\tau_0 - 0.006328\overline{y}}{\tau_0})^{4.4} , \qquad (3.7.3)$$

and

$\overline{y}=(2/3)y_m$, for field artillery,

$\overline{y}=(1/2)y_m$, for anti-aircraft artillery (up-hill, or down-hill shooting), where y_m is the maximum height of the trajectory.

The system of differential equations (3.7.1), is valid for any G-function of resistance that has the form:

$$G_D(v) = \begin{cases} A \cdot v^2 & for \quad v \le 256m/s \\ E \cdot (v - F/E) & for \quad v > 256m/s \end{cases} \qquad (3.7.4)$$

as long as the projectile speed satisfies the condition: $v > 256m/s$.

In case the ballistics temperature, ballistics density, and ballistics pressure are known (given by a met bulletin) we substitute in (3.7.3)

$$h_0(y) = \rho_B / \rho_0 . \qquad (3.7.5)$$

The Trajectory Equation of Low-Flight Projectiles

The equation of a projectile trajectory is a function of the horizontal range x, i.e., $y = f(x)$. Expanding $f(x)$ in Maclauren series, we can write approximately

$$y = f(0) + f'(0) \cdot x + \frac{f''(0)}{2!} \cdot x^2 + \frac{f'''(0)}{3!} x^3 + \frac{f^{IV}(0)}{4!} . \qquad (3.7.6)$$

Using the system of equations (3.7.1), we find the values of the derivatives at the departing point of the projectile:

$$f(0) = y_0 = 0, \qquad f'(0) = [\frac{dy}{dx}]_{x_0} = p_0 ,$$

$$f''(0) = [\frac{d^2 y}{dx^2}]_{x=0} = [\frac{dp}{dx}]_{x=0} = [-\frac{g}{v_x^2}]_{x=0} = -\frac{g}{v_0^2} ,$$

$$f'''(0) = \frac{d(-g/v_x^2)}{dx} = \frac{2g \cdot (dv_x / dx)_{x=0}}{v_{0x}^3} = -2gb \cdot \frac{v_{0x} - F/E}{v_{0x}^4} = -2gb \cdot \frac{v_0 - F/E}{v_0^4} ,$$

$$f^{IV}(0) = [d(-2gb \cdot \frac{v_x - F/E}{v_x^4})/dx]_{x=0} = -6gb^2 \cdot (v_0 - F/E) \cdot \frac{v_0 - (4/3) \cdot (F/E)}{v_0^6} .$$

Using the above values of the derivatives, and considering the fourth equation of (3.7.1), we obtain the equation of the projectile trajectory for relatively low trajectories:

$$y = \tan \alpha_0 \cdot x - \frac{g}{2v_0^2} \cdot x^2 - \frac{gb(v_0 - F/E)}{3v_0^4} \cdot x^3 -$$

$$- \frac{g \cdot b^2 (v_0 - F/E) \cdot [v_0 - (4/3) \cdot (F/E)]}{4v_0^6} \cdot x^4 . \qquad (3.7.7)$$

Ignoring the last term as very small, we have a less accurate but still satisfying equation for the projectile trajectory:

$$y = \tan \alpha_0 \cdot x - \frac{g}{2v_0^2} \cdot x^2 - \frac{g \cdot b \cdot (v_0 - F/E)}{3v_0^4} \cdot x^3 . \qquad (3.7.8)$$

Differentiating "y" and employing the fourth equation of (3.7.1), we express "$\tan \alpha$" as a function of the x-coordinate:

$$\tan \alpha = \tan \alpha_0 - \frac{g}{v_0^2} \cdot x - \frac{gb(v_0 - F/E)}{v_0^4} \cdot x^2 -$$

$$- \frac{g \cdot b^2 (v_0 - F/E) \cdot [v_0 - (4/3) \cdot (F/E)]}{v_0^6} \cdot x^3 , \qquad (3.7.9)$$

The last equation can be approximated and written in the form of the less accurate equation:

$$\tan\alpha = \tan\alpha_0 - \frac{g}{v_0^2}\cdot x - \frac{g\cdot b\cdot(v_0 - F/E)}{v_0^4}\cdot x^2. \qquad (3.7.10)$$

We can use the equation of the projectile flight, (3.7.7) or (3.7.8), for low trajectories to solve the following main problems of exterior ballistics:

- Find the range x_T when v_0, α_0, c are known.
- Find the launching angle α_0 when v_0, x, c are known.
- Find the ballistics coefficient c when v_0, α_0, and the coordinates of the impact point x_T and y_T are known.

To estimate the elements of a projectile trajectory, we can combine the above equations with the equations obtained in section 5.1, i.e., with equations:

$$x = 240t + \frac{(v_{x0} - F/E)}{b}(1 - e^{-b\cdot t}), \qquad (3.7.11)$$

$$v_x = F/E + (v_{x0} - F/E)\cdot e^{-b\cdot t}, \qquad (3.7.12)$$

$$\tan\alpha = \tan\alpha_0 - \frac{g}{(F/E)\cdot b}\ln(\frac{(F/E)\cdot e^{b\cdot t} + v_{x0} - F/E}{v_{x0}}) \qquad (3.7.13)$$

and

$$v_y = p\cdot v_x, \qquad (3.7.14)$$

where $p = \tan\alpha$, $p_0 = \tan\alpha_0$.

Note that the low-flight projectiles are usually launched at narrow angles.

The Projectile Drop

For any value of the x-coordinate, the drop of a projectile under the line of projectile departure is the difference $(x\tan\alpha_0 - y)$ that can be estimated by the equation (3.7.7) or (3.7.8), and can be presented in the following forms:

$$(x \tan \alpha_0 - y) = \frac{g}{2v_0^2} \cdot x^2 + \frac{gb(v_0 - F/E)}{3v_0^4} \cdot x^3 +$$
$$+ \frac{g \cdot b^2 (v_0 - F/E) \cdot [v_0 - (4/3) \cdot (F/E)]}{4v_0^6} \cdot x^4 , \qquad (3.7.15)$$

or

$$(x \tan \alpha_0 - y) = \frac{g}{2v_0^2} \cdot x^2 + \frac{g \cdot b \cdot (v_0 - F/E)}{3v_0^4} \cdot x^3 . \qquad (3.7.16)$$

The drop of the projectile fired with initial angle $\alpha_0 = 0°$ is

$$y = -\frac{g}{2v_0^2} \cdot x^2 - \frac{gb(v_0 - F/E)}{3v_0^4} \cdot x^3 -$$
$$- \frac{g \cdot b^2 (v_0 - F/E) \cdot [v_0 - (4/3) \cdot (F/E)]}{4v_0^6} \cdot x^4 , \qquad (3.7.17)$$

or

$$y = -\frac{g}{v_0^2} \cdot \frac{x^2}{2} - \frac{g \cdot b \cdot (v_0 - F/E)}{v_0^4} \cdot \frac{x^3}{3} . \qquad (3.7.18)$$

Note: that the actual method is not as accurate as the method we showed in preceding section. It gives good results for short distances.

Example 7.1 Antitank Fire

Find the departure angle α_0 needed to hit a tank that is at a horizontal range of 1,500 meters from a 45mm antitank cannon, if a projectile of with a mass 0.98 kg is fired with initial speed 1200 m/s. The ballistics coefficient of projectile with respect to the G_2-function (ICAO atmosphere) of resistance is 1.3965.

The temperature of the air is 25 degrees and the pressure is 755mm Hg. The humidity is 0%.

Solution:

For G_2-function of resistance $E = 0.143353$, $F = 30.2415$, $E / F = 210.96$.

Since the launching angle is relatively small, we assume $h(y) \approx h(\overline{y}) = 1$ and $v_{x0} = v_0 \cos\alpha_0 \approx v_0 = 1200 \, \text{m/s}$.

$$\overline{c} = \frac{p_0}{p_{0N}} \sqrt{\frac{\tau_{0N}}{\tau_0}} c = \frac{755}{760} \cdot \sqrt{\frac{288.15}{298.15}} \cdot 1.3965 = 1.3849 \,.$$

$$b = E \cdot \overline{c} \cdot h_0(\overline{y}) = (0.143353) \cdot (1.3849) \cdot (1) = 0.19853$$

Launching Angle

To find the launching angle, we substitute $y=0$, in the equation of the projectile trajectory (3.7.7), and all the other values.

$$\tan\alpha_0 \cdot x - \frac{g}{2v_0^2} \cdot x^2 - \frac{gb(v_0 - F/E)}{3v_0^4} \cdot x^3 -$$

$$- \frac{g \cdot b^2 (v_0 - F/E) \cdot [v_0 - (4/3) \cdot (F/E)]}{4v_0^6} \cdot x^4 =$$

$$= \tan(\alpha_0) \cdot (2800) - \frac{g}{2 \cdot (1200)^2} \cdot (2800)^2 - \frac{g \cdot (0.19853) \cdot (1200 - 210.96)}{3 \cdot (1200)^4} (2800)^3 -$$

$$- \frac{g \cdot (0.19853)^2 \cdot (1200 - 210.96) \cdot [(1200 - (4/3) \cdot (210.96)]}{4 \cdot (1200)^6} \cdot (2800)^4 = 0.$$

Hence, we find that the departure angle is

$$\alpha_0 = \arctan(0.01203) = 0.6892° \,.$$

3.8. Parametric Equations of the Inclined Trajectory

In section 2.23, we have introduced a new method to determine the departure angle or the angle of projection needed to hit a target located at an inclined plane based on the standard range tables.

Now we will see an approximate method to determine the elements of the trajectory of a projectile when the target is located on an inclined plane.

We generalize the formulae obtained in section 3.4 of the *"Exterior Ballistics with Applications"*, solving the differential equations of projectile flight by formally replacing the Siacci's approximate function of resistance,

$$K_D(v) = \begin{cases} 1.212 \cdot 10^{-4} v^2 & for \quad v \le 256m/s \\ (v - 240)/3 & for \quad v > 256m/s \end{cases},$$

with the a G-function of resistance,

$$G_D(v) = \begin{cases} A \cdot v^2 & for \quad v \le 256m/s \\ E \cdot v - F & for \quad v > 256m/s \end{cases}. \qquad (3.8.1)$$

In analogy with the method used in *"Exterior Ballistics with Applications"* to solve the differential equations of flight, we find the solution of the differential equations (5.1.1),

$$\begin{cases} \dfrac{dv_x}{dt} = -\bar{c} \cdot h_0(y) \cdot G_D(v) \cdot \dfrac{v_x}{v} \\[2mm] \dfrac{dp}{dt} = -\dfrac{g}{v_x} \\[2mm] \dfrac{dx}{dt} = v_x \\[2mm] \dfrac{dy}{dt} = v_y \end{cases} \qquad (3.8.2)$$

We will write differential equations (3.8.2) in the system of curvilinear coordinates $\bar{x}$, $\bar{y}$, (figure 9).

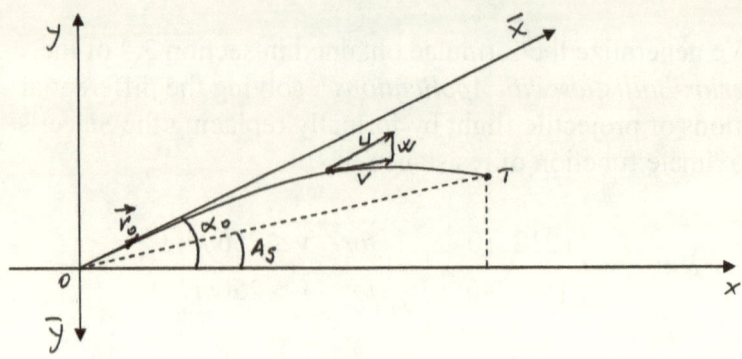

Figure 9

The coordinates (x, y) of the projectile in the rectangular Cartesian coordinate system and the coordinates $\bar{x}, \bar{y}$ of the projectile in the curvilinear system are related by the relations

$$x = \bar{x}\cos\alpha_0, \quad y = \bar{x}\sin\alpha_0 - \bar{y} \qquad (3.8.2)$$

The equations (3.8.2) in the curvilinear system of coordinates can be presented in the form:

$$\begin{cases} \dfrac{du}{dt} = -\bar{c}\cdot h_0(y)G_D(v)\dfrac{u}{v} \\[2mm] \dfrac{d\bar{p}}{dt} = \dfrac{g}{u} \\[2mm] \dfrac{d\bar{x}}{dt} = u \\[2mm] \dfrac{d\bar{y}}{dt} = w \end{cases} , \qquad (3.8.3)$$

where

$$\vec{v} = \vec{u} + \vec{w}, \quad \bar{p} = w/u, \quad v^2 = u^2 + w^2 - 2u\cdot w\cdot\sin\alpha_0. \qquad (3.8.4)$$

($\vec{u}$ and $\vec{w}$ are the components of the velocity $\vec{v}$ in the curvilinear coordinate system.)

From the triangle of velocities, presented in figure 10, we have [66]

$$u = v\frac{\cos\alpha}{\cos\alpha_0} \tag{3.8.5}$$

$$\begin{cases} \dfrac{du}{dt} = -\overline{c}\cdot h_0(y)\dfrac{\cos\alpha_0}{\cos\alpha}G_D(u\dfrac{\cos\alpha_0}{\cos\alpha}) \\ \dfrac{d\overline{p}}{dt} = \dfrac{g}{u} \\ \dfrac{d\overline{x}}{dt} = u \\ \dfrac{d\overline{y}}{dt} = w \end{cases} . \tag{3.8.5}$$

The system of differential equations (3.8.5) can be solved numerically.

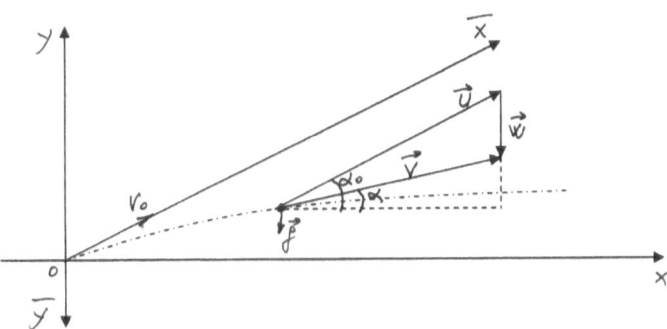

Figure 10

Approximate Analytical Solutions

The system of differential equations in curvilinear coordinates, (3.8.5) for projectile speed $v > 256m/s$ can be solved in quadratures when the distance to the target, $D = oT$, is relatively short.

---

66. The speed u given by (5.3.5) is the Siacci pseudo-speed that is introduced in Chapter 5 of *"Exterior Ballistics with Applications"*.

For such inclined ranges, the projectile trajectory deviates slightly from the direction of the initial velocity $\vec{v}_0$ (direction of $\bar{x}$-axis). We can assume that at any point $P(\bar{x}, \bar{y}) = P(x, y)$ in the trajectory of flight, the speed of the projectile is approximately equal to the component u of the velocity, i.e.,

$$\vec{v} \approx \vec{u}, \quad v \approx u . \tag{3.8.6}$$

The G-function of resistance can be approximated and written in the form:

$$G_D(v) = \begin{cases} A \cdot u^2 & for \quad u \le 256m/s \\ E \cdot (u - F/E) & for \quad u > 256m/s \end{cases} . \tag{3.8.7}$$

To compensate somehow the error we make using approximation (3.8.6), we multiply the right side of the first equation of (3.8.5) the quantity ($\cos\alpha_0/\cos\alpha$) that is less than one. [67]. Thus, for speed $v > 256m/s$, the system of differential equations (3.8.5) can be written:

$$\begin{cases} \dfrac{du}{dt} = -\bar{c} \cdot E \cdot h_0(y)(\bar{u} - F/E) \\[2mm] \dfrac{d\bar{p}}{dt} = \dfrac{g}{u} \\[2mm] \dfrac{d\bar{x}}{dt} = u \\[2mm] \dfrac{d\bar{y}}{dt} = w \end{cases} \tag{3.8.8}$$

Now, we replace the density function $h_0(y)$ with a hypothetical constant density value $h_0(\hat{y})$ that have the same effect on the trajectory of flight as the density function itself.

For the inclined fire, we consider that

$$h_0(y) = h_0(\hat{y}), \tag{3.8.9}$$

---

67. Siacci uses a compensation variable factor $\cos^2\alpha_0/\cos\alpha$ instead of the factor $\cos\alpha_0/\cos\alpha$.

where

$$h_0(\hat{y}) = (\frac{\tau_0 - 0.006328\hat{y}}{\tau_0})^{4.4}, \quad \hat{y}=(1/2)y_{max},$$
(3.8.10)

y_{max} is the maximum height of the projectile trajectory.

Denoting

$$B = \bar{c} \cdot E \cdot h(\hat{y}),$$
(3.8.11)

and substituting in (3.8.8), we obtain the system of differential equations in a simplified form:

$$
\begin{cases}
\dfrac{du}{dt} = -B(u - F/E) \\
\dfrac{d\bar{p}}{dt} = -(-\dfrac{g}{u}) \\
\dfrac{d\bar{x}}{dt} = u \\
\dfrac{d\bar{y}}{dt} = w
\end{cases}
.
$$
(3.8.12)

System (3.8.12) has an identical form as system (3.6.11) section 3.6.

The solution of (3.8.12) is obtained by formally substituting the curvilinear variables in the solutions of (3.6.11).

Equations of Projectile Trajectory in Curvilinear System

Solving (3.8.12), we obtain:

- the parametric equations of the projectile trajectory

$$\bar{x} = (F/E)t + \frac{(v_0 - F/E)}{B}(1 - e^{-B \cdot t}),$$
(3.8.13)

and

$$\bar{y} = \frac{g}{(F/E) \cdot B} \int_0^{t}(F/E + (v_0 - (F/E) \cdot e^{-B \cdot t}) \cdot \ln(\frac{(F/E)e^{B \cdot t} + v_0 - (F/E)}{v_0})dt$$
(3.8.14)

- the value of the parameter $\bar{p} = w / u$:

$$\bar{p} = \frac{g}{(F / E) \cdot B} \ln(\frac{(F / E) \cdot e^{B \cdot t} + v_0 - F / E}{v_0}), \qquad (3.8.15)$$

- the components of speed in curvilinear coordinates:

$$u = F / E + (v_0 - F / E) \cdot e^{-B \cdot t}, \qquad (3.8.16)$$

and

$$w = \bar{p} \cdot u \qquad (3.8.17)$$

Note that: $\bar{p}_0 = w_0 / u_0 \approx 0$, $\bar{p} = w / u \geq 0$.

Elements of the Trajectory in Rectangular Cartesian System

Once the elements of the trajectory in curvilinear coordinates are calculated using the above formulas (3.8.13)-(3.8.17), we are able to find the elements of the trajectory in rectangular Cartesian coordinates at any time during the flight.

Thus, from (3.8.2), we find

Cartesian Coordinates of the Projectile

$$x = \bar{x} \cdot \cos\alpha_0, \quad y = \bar{x}\sin\alpha_0 - \bar{y} \quad (\text{or: } y = x\tan\alpha_0 - \bar{y}). \qquad (3.8.18)$$

The Projectile Speed

$$v^2 = u^2 + w^2 - 2u \cdot w \cdot \sin\alpha_0 \quad (\text{or: } v \approx u) \qquad (3.8.19)$$

The Angle of Flight

$$\tan\alpha = \frac{\bar{p} - \sin\alpha_0}{\cos\alpha_0}. \qquad (3.8.20)$$

The relation (3.8.20) can be easily found using the equation we get from the geometry of figure 10 applying the sine theorem:

$$\overline{p} = \frac{\sin(\alpha_0 + \alpha_T)}{\cos \alpha_T}. \tag{3.8.21}$$

Slant Range

The relation between the slant range D and the $\overline{x}$-coordinate of projectile is

$$D = \frac{\cos \alpha_0}{\cos A_S} \overline{x}. \tag{3.8.22}$$

The Equations of Projectile Trajectory in Curvilinear System

Substituting (3.8.2) into (3.8.13) and (3.8.14), we obtain the equations of the projectile trajectory in rectangular Cartesian system:

$$x = [(F/E) \cdot t + \frac{(v_0 - F/E)}{B} \cdot (1 - e^{-B \cdot t})] \cdot \cos \alpha_0 \tag{3.8.23}$$

$$y = x \tan \alpha_0 -$$

$$- \frac{g}{(F/E) \cdot B} \int_0^t (F/E + (v_0 - (F/E) \cdot e^{-B \cdot t}) \cdot \ln(\frac{(F/E)e^{B \cdot t} + v_0 - (F/E)}{v_0}) dt ,$$

$$\tag{3.8.24}$$

Note that:

the value of $\overline{y}$, is practically the "drop" of the projectile, from the direction of departure angle ($\overline{x}$-axis), during the time of flight.

3.9. Equation of Inclined Trajectory in Cartesian Coordinates

The equations of the projectile trajectory in curvilinear coordinates (3.8.13) and (3.8.14) as well as the equations of the other obtained formulae, can be used to find the elements of the trajectory for a given departure angle, initial speed, and ballistics coefficient at any time during the flight.

The reverse problem, i.e., the problem that requires finding the angle of departure needed to hit a target located in the inclined plane, is more difficult to be solved using those equations.

Anyway, for the projectiles flying with speeds greater than 256 m/s, we can find approximate equations for inclined trajectory in rectangular Cartesian coordinates.

We express the system of differential equations (3.8.12) through the curvilinear $\bar{x}$ -coordinate:

$$
\begin{cases}
\dfrac{du}{d\bar{x}} = -B\dfrac{(u - F/E)}{u} \\[2mm]
\dfrac{d\bar{p}}{d\bar{x}} = -(-\dfrac{g}{u^2}) \\[2mm]
\dfrac{dt}{d\bar{x}} = \dfrac{1}{u} \\[2mm]
\dfrac{d\bar{y}}{d\bar{x}} = p
\end{cases}
\qquad (3.9.1)
$$

where

$$
\bar{x} = x / \cos\alpha_0, \quad \bar{y} = x \tan\alpha_0 - y. \qquad (3.9.2)
$$

Since the system of the differential equations (3.9.1) has identical form as the system (5.2.1), we can easilyy obtain the following equations by formally replacing curvilinear coordinates for the Cartesian coordinates.

Equation of Projectile Trajectory in Curvilinear Coordinates

The equation of projectile trajectory in curvilinear coordinates is

$$
\bar{y} = \frac{g}{2u_0^2} \cdot \bar{x}^2 + \frac{g \cdot B(u_0 - F/E)}{3u_0^4} \cdot \bar{x}^3 +
$$
$$
+ \frac{g \cdot B^2 (u_0 - F/E) \cdot [u_0 - (4/3) \cdot (F/E)]}{4u_0^6} \cdot \bar{x}^4. \qquad (3.9.3)
$$

Neglecting the last term, we obtain another approximate formula, but less accurate than (3.9.3),

$$\bar{y} = \frac{g}{2u_0^2} \cdot \bar{x}^2 + \frac{g \cdot B(u_0 - F/E)}{3u_0^4} \cdot \bar{x}^3 . \qquad (3.9.4)$$

Parameter $\bar{p}$

Differentiating $\bar{y}$ with respect to $\bar{x}$ and employing the fourth equation of (3.9.1) we find $\bar{p}$ as a function of $\bar{x}$:

$$\bar{p} = \frac{g}{u_0^2} \cdot \bar{x} + \frac{gB(u_0 - F/E)}{u_0^4}\bar{x}^2 + gB^2(u_0 - 240) \cdot \frac{u_0 - (3/4) \cdot (F/E)}{u_0^6}\bar{x}^3 .$$

$$(3.9.5)$$

or, less accurately,

$$\bar{p} = \frac{g}{u_0^2} \cdot \bar{x} + \frac{gB(u_0 - F/E)}{u_0^4}\bar{x}^2 \qquad (3.9.6)$$

The Equation of the Inclined Trajectory in Cartesian Coordinates

Employing (3.9.2), from the equation of the trajectory (3.9.3), we obtain the equation of the inclined trajectory in Cartesian coordinates:

$$y = x \tan\alpha_0 - \frac{g}{2v_0^2 \cdot \cos^2\alpha_0} \cdot x^2 - \frac{g \cdot B(v_0 - F/E)}{3v_0^4 \cos^3\alpha_0} \cdot x^3 -$$

$$- \frac{g \cdot B^2(v_0 - F/E) \cdot [v_0 - (4/3) \cdot (F/E)]}{4v_0^6 \cos^4\alpha_0} \cdot x^4 . \qquad (3.9.7)$$

The Projectile Speed

$$v^2 = u^2 + w^2 - 2u \cdot w \cdot \sin\alpha_0 \qquad (3.9.8)$$

The Angle of Flight

$$\tan\alpha = \frac{\bar{p} - \sin\alpha_0}{\cos\alpha_0} . \qquad (3.9.9)$$

where

$$\bar{p} = \frac{\sin(\alpha_0 + \alpha_T)}{\cos\alpha_T} , \qquad (3.9.10)$$

The formulas obtained in this chapter can be used to solve the two main problems of exterior ballistics:

- Find the slant range $\overline{D}$ when are known v_0, α_0, c are known.
- Find launching angle α_0 when are known $v_0, \overline{D}, c$ are known.

We can employ the equations obtained above together with the equations of section 5.3 to solve different problems of related with the inclined firing.

The Projectile Drop in Horizontal Fire

A projectile fired horizontally drops continuously under the horizontal line that has the direction of the launching speed (figure 11).

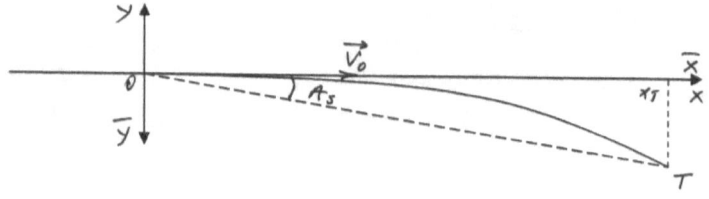

Figure 11

Since the $\overline{x}$-axis coincides with x-axis, we obtain the drop of the substituting "x" for "$\overline{x}$" in (3.9.2) or (3.9.3) and $y = -\overline{y}$.

Thus, for the drop of the projectile in Cartesian coordinates, we obtain the following equation:

$$y = -\frac{g}{2v_0^2} \cdot x^2 - \frac{g \cdot B(v_0 - F/E)}{3v_0^4} \cdot x^3 -$$

$$- \frac{g \cdot B^2(v_0 - F/E) \cdot [v_0 - (4/3) \cdot (F/E)]}{4v_0^6} \cdot x^4. \qquad (3.9.11)$$

Example 9.1

For a 14.5mm bullet of an antiaircraft weapon, find the projection angle A and the time of flight "t" to the target when the firing distance and the angle of elevation of the target are respectively $D=1000m$

and $E=60°$. The initial speed of the bullet is $v_0=945\,\text{m/s}$, while the ballistic coefficient is $c=1.64$.

Solution:

Estimate B.

As a first approach, we consider

$$\bar{x} \approx \overline{D} = 1000m ,$$

$$\bar{y}=y_m/2, \text{ where } y_m = D\sin A = (1000)\cdot \sin(60) = 866.03m .$$

$$h(\hat{y}) = (\frac{289.08 - 0.006328\cdot (866.03)}{289.08})^{4.4} = 0.9192,$$

and

$$B = \frac{c\cdot h(\hat{y})}{3} = \frac{(1.62)(0.9192)}{3} = 0.496368,$$

Substituting in (27),

$$\bar{x} = 240t + \frac{v_0 - 240}{B}(1 - e^{-B\cdot t}),$$

we have

$$1000 = 240t + \frac{(945 - 240)}{0.4964}(1 - e^{-0.4964\cdot t})$$

Simplifying, we can write

$$0.70413 - 0.16899t = +(1 - e^{-0.4964\cdot t})$$

Solving the above exponential equation for t (using a graphing calculator TI83+) we find the time of flight $t=1.321s$.

We find

$$\bar{y}=-\frac{g}{240\cdot B}\int_0^t (240+(v_0-240)\cdot e^{-B\cdot t})\cdot\ln(\frac{240e^{B\cdot t}+v_0-240}{v_0})dt=$$

$$=-0.0823\int_0^{1.321}(240+705e^{-0.4964\cdot t})\cdot\ln(0.253968e^{0.4964\cdot t}+1.09302)dt$$

The integral on the right side of the above equation has the value 90.135. Thus, the "drop" is

$$\bar{y}=0.0823(90.135)=7.42 \text{ m.}$$

Substituting in (32), we find that,

$$A=\frac{\bar{y}}{D}\cos(A)=\frac{7.42}{1000}\cos(60°)=0.00376 \text{ radian.}$$

or

$$A=0.00376(180/\pi)=0.2154°=12.93'$$

Using (37), we find that the curvilinear coordinate $\bar{x}$ is

$$\bar{x}=D\frac{\cos(E)}{\cos(E+A)}=1,000\frac{\cos(60°)}{\cos(60.2154°)}=1006.56 \text{ m}$$

Improving Results

We now consider $\bar{x}=1006.56$. Repeating the same operations for that value of $\bar{x}$, we find

$$t=1.332s, \quad \bar{y}=7.53m, \quad A=0.21585°=12.95'$$

projectile speed $v_0=945m/s$,   ballistics coefficient $c=1.64$.

Results:

Launching angle is $\alpha_0=60.22168°$, time of flight is $t=1.355s$, impact speed is $v_T=579.06m/s$, impact angle $\alpha_T=59.69°$

Hence, we find that the aiming angle is

$$A=\alpha_0-E=60.22168°-60°=0.22168°=13.3'$$

4

MODERN APPROACH TO THE THEORY OF CORRECTIONS

Introduction

The elements of the trajectory of the projectile flight presented in the firing range tables of a projectile are calculated for the standard atmosphere conditions and for some standard ballistics characteristics of the projectile, including the standard departure speed.

Since in practice, the meteorological characteristics of the atmosphere and the projectile ballistics characteristics are usually different from the respective characteristics of the standard atmosphere and standard ballistics characteristics of the projectile, in general, in range tables are included the so-called "range corrections".

Range corrections are used to correct the departure angle, when shooting is performed in non-standard atmosphere at the sea level or over it.

In the firing range tables are shown the range corrections that are related to small deviations in temperature of air and pressure as well as the range corrections related with small changes in range-wind, and cross-wind.

Another category of range corrections is related with small changes in projectile departure speed and departure angle.

In practice, the departure speed of a projectile can somewhat change:

- as result of consumption of the barrel of the firearm and the increase in chamber capacity,
- as result of variations in projectile mass within the manufacturing tolerances (for example, different lots of cannon projectiles have different masses),
- as result of change in ballistics characteristics of the propellant mainly due to the changes in temperature, mass, and variations in their shape and size (which reduces/increases the initial burning surface).

The numerical methods and the PC programs that are in use to solve the differential equations of the projectile flight can predict the trajectory of the projectile flight with acceptable accuracy even when the projectile ballistics characteristics and the characteristics of the atmosphere are not standard.

The range tables of a cannon perform the same tasks in gunnery, but in them, there are included the correction quantities necessary to fire in non standard firing conditions.

The accuracy of the corrections depends on the method and the G-function of resistance that we employ to estimate those corrections as well as in the accuracy of range tables.

The correction gives acceptable results for "small" changes in the projectile speed, departure angle, and ballistics coefficient, which are within the following limits [68]:

$$\Delta v_0 = \pm 25 m/s, \ \Delta \alpha_0 = \pm 1.685°, \ \Delta c = \pm 10\% \cdot c.$$

Though the range corrections can be calculated using numerical integration of the differential equations of flight, the theory of corrections is useful for theoretical and practical purposes.

---

68. Field Artillery, Volume 6, p. 134, DND Canada, 1992

4.1. Effects of Changes in Characteristics of Atmosphere

Hereafter, we show a new approach to determine the angle of departure for a projectile fired in non-standard atmosphere based on the standard range tables of a firearm, avoiding the use of corrections presented in range tables, or other traditional methods.

The method considers standard all the other values of the main factors that influence the projectile flight as, for example, the temperature of propellant charge of the projectile, the projectile departure speed, the projectile mass and diameter (or form coefficient), etc..

Correction for Changes in Pressure and Temperature

We consider only small changes in the standard values of the atmospheric pressure and the virtual temperature.

The deviation in pressure and temperature of air (from the standard values) produces the deviation of the projectile flight from the trajectory predicted for the standard atmosphere.

The elements of the new similitude projectile trajectory deviate as well from the corresponding elements of the projectile trajectory in standard atmosphere and can be calculated using (4.1.1)-(4.1.5).

The horizontal range depends on the air pressure p and the virtual temperature τ, i.e., we can write that $x_T = f(\tau, p)$. Assuming that the small changes in the variables p and τ are independent, the change Δx_T in the horizontal range x_T that corresponds to small changes Δp and $\Delta \tau$ (respectively in pressure and temperature), is

$$\Delta x_T = \frac{\partial x_T}{\partial p} \Delta p + \frac{\partial x_T}{\partial \tau} \Delta \tau . \qquad (4.1.1)$$

The departure angle is a function of the horizontal range x_T, i.e., $\alpha_0 = \alpha_0(x_T)$. A small change Δx_T in the horizontal range produces a small change in the departure angle

$$\Delta \alpha_0 = \frac{d\alpha_0}{dx_T} \cdot \Delta x_T . \qquad (4.1.2)$$

Hence, for the departure angle that corresponds to the actual value $(x_T + \Delta x_T)$ of the horizontal range, we can write:

$$\alpha_{02} = \alpha_0 + \frac{d\alpha_0}{dx_T} \cdot \Delta x_T. \tag{4.1.3}$$

Note that

$$\frac{\partial x_T}{\partial p}, \frac{\partial x_T}{\partial \tau}, . \tag{4.1.4}$$

are corrections coefficients related with small changes respectively in pressure, and temperature, while $\partial \alpha_0 / \partial x_T$ is the correction coefficient of the departure angle that results from a small change in horizontal range.

The above theoretical way shows that once we have found the change in horizontal range (4.1.1), we have to correct the departure angle using (4.1.2) to match the new horizontal range with the departure angle.

Trajectory-Similitude and Snell's Law Correction Method

We use the similitude formulae (2.7.29)-(2.7.34), to calculate the horizontal range X_T, the departure angle α_{02}, and other elements $(t_{T2}, \alpha_{T2}, Y_m, V_T)$ of the trajectory of a projectile flying in non-standard atmosphere using the range x_T, the departure angle α_0, and the other elements $(t_T, \alpha_T, y_m, v_T)$ of the similitude trajectory at the sea -level standard atmosphere, i.e., we employ the following formulae:

Horizontal range:

$$X_T = J \cdot x_T, \tag{4.1.5}$$

Departure angle:

$$\sin \alpha_{02} = J \sin \alpha_0, \text{ or } \alpha_{02} = \arcsin(J \cdot \sin \alpha_0), \tag{4.1.6}$$

Time of flight:

$$t_{T2} = J \cdot t_T, \tag{4.1.7}$$

Terminal angle:

$$\alpha_{T2} = ar\sin(J \cdot \sin \alpha_T). \tag{4.1.8}$$

Maximum altitude:

$$Y_m = J^2 y_m, \tag{4.1.9}$$

where

$$J = \frac{p_{0N}}{p_2} \sqrt{\frac{\tau_2}{\tau_{0N}}}, \tag{4.1.10}$$

while p_{0N}, τ_{0N} are respectively the virtual temperature and pressure for the standard atmosphere, while p_2, τ_2 are the virtual temperature and pressure of air in non standard atmosphere.

The initial and the terminal speed of the projectile flying in non-standard atmosphere are equal respectively to the initial speed of and the terminal speed of the same projectile flying in standard atmosphere, i.e.,

$$V_T = v_T, \quad V_0 = v_0. \tag{4.1.11}$$

The formulas we use to find the correction in angle are (figure 12.):

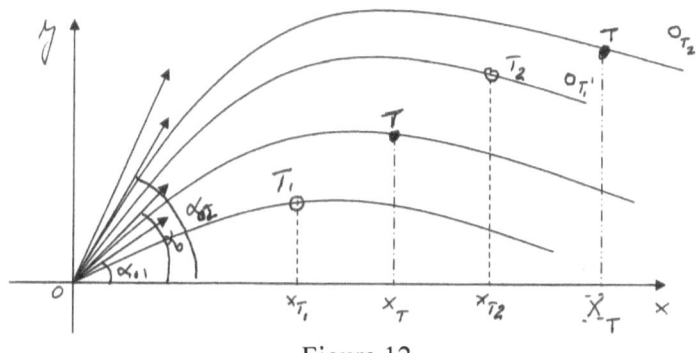

Figure 12

Correction in Departure Angle:

$$\Delta \alpha = \frac{\alpha_2 - \alpha_1}{X_{T2} - X_{T1}} \cdot h. \tag{4.1.12}$$

"Corrected" Departure Angle:

$$\alpha_{02} = \alpha_{01} + \frac{\alpha_2 - \alpha_1}{X_{T2} - X_{T1}} \cdot h.$$ (4.1.13)

where

$$\alpha_1 = \alpha_1(X_{T1}), \ \alpha_2 = \alpha_2(X_{T2}), \ h = X_T - X_{T1}, \text{ or } h = X_T - X_{T2}.$$
(4.1.14)

To "correct" the departure angle already known for a standard atmosphere, i.e., to find the angle needed to hit the target in the actual non-standard atmosphere, we use the procedure illustrated in examples 1.1-1.3.

The example 1 and example 2 demonstrates the method we use to find the correction in the departure angle that corresponds respectively:
- only to a change in pressure of $\Delta p = p_2 - p_{0N} = \pm 10mm \ Hg.$,
- only to a change in temperature of $\Delta T = T_2 - T_{0N} = \pm 10° C$.

The example 1.3 illustrates the way we correct the departure angle when both temperature and pressure deviates from their respective standard values.

The linear approximations are acceptable for small changes in the characteristics of atmosphere, but they give large errors for relatively large changes in the characteristics of atmosphere.

Example 1.1 Correction for a Small Change in Air Pressure

The elements of the trajectory of the a 122mm projectile of the Russian cannon 122mm Model 1060, launched at a speed of 885m/s, for the horizontal range $x_T = 12000m$ meters, in TSA atmosphere (virtual temperature $\tau_{0N} = 289.08° K$ ([temperature of dry air 15 degrees Celsius, relative humidity of air 50%, pressure of water vapors, 6.35mm Hg.)]), are :

departure angle $\alpha_{01} = 8.6°$, time of flight to the target $t_T = 22s$, terminal angle $\alpha_T = 16°$, and trajectory maximum altitude $y_m = 615m$.

Find the departure angle needed to hit the target located at the horizontal range 12,000 meters if there is:

(a) a change of +10 mm Hg in atmospheric pressure.
(b) a change of -10 mm Hg in atmospheric pressure.

Solution:

(a) The air pressure changes from 750mm Hg to 760mm Hg.

$$\Delta p = p_2 - p_{0N} = 760 - 750 = +10mm\ Hg.$$

The temperature and the humidity of air remain unchanged.
We have:

$$J_1 = \frac{p_{0N}}{p_2}\sqrt{\frac{\tau_2}{\tau_{0N}}} = \frac{750}{760}\sqrt{\frac{289.08}{289.08}} = 0.9868.$$

The change in atmospheric pressure produces a change in the corresponding virtual temperature (that is negligible). Indeed, the virtual temperature becomes

$$\tau_2 = \frac{T_2}{(1 - 0.3785 \cdot e_2 / p_2)} = \frac{288.15}{(1 - 0.3785 \cdot 6.35 / 760)} = 289.64°\ K.$$

We have practically the same value:

$$J_1 = \frac{p_{0N}}{p_2}\sqrt{\frac{\tau_2}{\tau_{0N}}} = \frac{750}{760}\sqrt{\frac{289.06}{289.08}} \approx \frac{750}{760} = 0.9868.$$

The horizontal range of the projectile in non-standard atmosphere, for the corresponding similitude trajectory, is

$$X_{T1} = J_1 x_T = 0.9868 \cdot (12000) = 11841.6m.$$

The corresponding departure angle is

$$\alpha_1 = ar\sin(J_1 \sin\alpha_{01}) = \arcsin(0.9868 \cdot \sin(8.6°)) = 8.48564°.$$

Now consider that the air pressure change is "-10mm Hg.", and it changes from 750mm Hg to 740mm Hg. We have:

$$J_2 = \frac{p_{0N}}{p_2}\sqrt{\frac{\tau_2}{\tau_{0N}}} = \frac{750}{740}\sqrt{\frac{289.08}{289.08}} = 1.01351° \, K,$$

$$X_{T2} = J_2 x_T = 1.01351 \cdot (12000) = 12162.16 m$$

and

$$\alpha_2 = ar\sin(J\sin\alpha_{01}) = \arcsin(1.01351 \cdot \sin(8.6°)) = 8.71712°.$$

The change in range,

$$X_{T2} - X_{T1} = (12162.16 - 11841.6) = 320.56 m,$$

corresponds to a change in departure angle of

$$\alpha_2 - \alpha_1 = (8.71712° - 8.4943°) = 0.22282°.$$

When the pressure is 760mm (increase: +10mm), the projectile will get to the target if the change in angle will be proportional to the change in range

$$h = X_T - X_{T1} = (12000 - 11841.6) = 158.4 m.$$

Assuming that the change in angle is proportional to the change in range, by interpolation, we find that the departure angle $\alpha_0 = 8.6°$ should be increased by the quantity:

$$\Delta\alpha_0 = (\frac{0.22282°}{320.56}) \cdot 158.4 = 0.1101.$$

Thus the "corrected" firing angle needed to hit the target is

$$\alpha_{02} = \alpha_0 + \Delta\alpha_0 = 8.6° + 0.1101° = 8.7101°.$$

(b) The air pressure changes from 750mm Hg to 740mm Hg., i.e.,

$$\Delta p = p_2 - p_{0N} = 740 - 750 = -10mm\ Hg.$$

The corresponding departure angle is

$$\alpha_{02} = \alpha_0 + \frac{\alpha_2 - \alpha_1}{X_{T2} - X_{T1}} \cdot h =$$

$$= 8.6 + \frac{(8.71712° - 8.48564)}{(12162.16 - 11841.6)} \cdot (12000 - 12162.16) = 8.48290°$$

The angle correction is

$$\Delta\alpha_0 = \frac{\alpha_2 - \alpha_1}{X_{T2} - X_{T1}} \cdot h = -0.1171°.$$

In the following table there are given the data obtained using the above method as well as the data calculated using the range table of the 162mm cannon 162mm and the data calculated using the PC program Aca122.Bas.

Table 1: Range 12,000 meters

	Pressure 760 mm Hg		Pressure 740 mm Hg	
	Angle	Correction	Angle	Correction
Modern Method	8.7101	+ 0.1101	8.4829	- 0.1171
Range Table	8.7050	+ 0.1050	8.4950	- 0.1050
ACA122.Bas	8.7008	+ 0.1008	8.4980	-0.1020

Example 1.2 Correction for a Small Change in Air Temperature of air

The elements of the trajectory of the projectile 122mm of the Russian 122mm cannon 122mm (launched at a speed of 885m/s) for the horizontal range $x_T = 16000m$ meters, in TSA atmosphere (virtual temperature $\tau_{0N} = 289.08° K$, temperature of dry air 15 degrees Celsius, relative humidity of air 50%, pressure of water vapors, 6.35mm Hg) are :

departure angle $\alpha_{01} = 15.70°$;, time of flight to the target $t_T = 36s$;, terminal angle $\alpha_T = 31°$;, trajectory vertex altitude $y_m = 1660m$.

Find the departure angle needed to hit the target at the horizontal range of 16,000 meters if there is:

(a) a change of +10 degree Celsius in the temperature of air.
(b) a change of -10 degree Celsius in the temperature of air.

Solution:

(a) Change of +10 degree Celsius in the air temperature of air

$$\Delta T = T_2 - T_{0N} = 298.15 - 288.15 = +10° C.$$

From table 2, section 2.2, we find that the pressure of water vapors is $e = 11.88 mmHg$. The virtual temperature is

$$\tau_2 = \frac{T_2}{(1 - 0.3785 \cdot e_2 / p_2)} = \frac{298.15}{(1 - 0.3785 \cdot 11.88 / 750)} = 299.95° K$$

We have

$$J_2 = \frac{p_{0N}}{p_2} \sqrt{\frac{\tau_2}{\tau_{0N}}} = \frac{750}{750} \sqrt{\frac{299.95}{289.08}} = 1.01868,$$

$$X_{T2} = J_2 \cdot x_T = 1.01868 \cdot (16000) = 16298.84m$$

and

$$\alpha_{02} = ar \sin(J \cdot \sin \alpha_{01}) = \arcsin(1.01868 \cdot \sin(15.70°)) = 16.0011°.$$

In a similar way, we find the data that correspond to a change -10 degree Celsius. We have:

$$\Delta T = T_2 - T_{0N} = 278.15 - 288.15 = -10° C,$$

$$\tau_2 = \frac{T_2}{(1 - 0.3785 \cdot e_2 / p_2)} = \frac{278.15}{(1 - 0.3785 \cdot (3.27) / (750))} = 278.61° K,$$

$$J_1 = \frac{p_{0N}}{p_2}\sqrt{\frac{\tau_2}{\tau_{0N}}} = \frac{750}{750}\sqrt{\frac{278.61}{289.08}} = 0.98172 \,,$$

$$X_{T1} = J_1 \cdot x_T = 0.98172 \cdot (16000) = 15707.58m$$

and

$$\alpha_{02} = ar\sin(J\sin\alpha_{01}) = \arcsin(0.98172 \cdot \sin(15.70°\,)) = 15.40587° \,.$$

The "corrected" departure angle needed to hit the target at the distance of 16,000 meters is

$$\alpha_{02} = \alpha_0 + \frac{\alpha_2 - \alpha_1}{X_{T2} - X_{T1}} \cdot h = 15.7 +$$

$$+ \frac{(16.00011° - 15.40581°\,)}{(16298.84 - 15707.58)} \cdot (16000 - 16298.84) = 15.3991°\,.$$

Correction in departure angle is

$$\Delta\alpha = \frac{\alpha_2 - \alpha_1}{X_{T2} - X_{T1}} \cdot h = -0.3009° \,.$$

(b) A change of -10 degree Celsius in the temperature of air

$$\Delta T = T_2 - T_{0N} = 278.15 - 288.15 = -10° \, C \,.$$

The "corrected" departure angle needed to hit the target at the distance of 16,000 meters is

$$\alpha_{02} = \alpha_0 + \frac{\alpha_2 - \alpha_1}{X_{T2} - X_{T1}} \cdot h =$$

$$= 15.7 + \frac{(16.00011° - 15.40581°\,)}{(16298.84 - 15707.58)} \cdot (16000 - 15707.58) = 15.9944°\,.$$

Correction in departure angle needed to hit the target is

$$\Delta\alpha_0 = \frac{\alpha_2 - \alpha_1}{x_2 - x_1} \cdot h = 0.29441°.$$

In the following table there are given the data obtained using the above method as well as the data calculated using the range table of 162mm cannon 162mm and the data calculated using the PC program Aca122.Bas.

Table 1: Range 16,000 meters

	25 degrees C		5 degrees C	
	Angle	Correction	Angle	Correction
Modern Method	15.3991	- 0.3009	15.9944	+ 0.2944
Range Table	15.1943	- 0.5057	16.2057	+ 0.5057
ACA122.Bas	15.4268	- 0.2732	15.9773	+ 0.2773

Example 1.3 Correction for Changes in Temperature and Pressure

The elements of the trajectory of the 122mm projectile 122mm of the Russian 122mm cannon 122mm (departure speed of 885m/s) for the horizontal range $x_T = 10000m$ meters in TSA atmosphere (virtual temperature $\tau_{0N} = 289.08° K$, temperature of dry air 15 degrees Celsius, relative humidity of air 50%, pressure of water vapors, 6.35mm Hg.) are:

departure angle, $\alpha_{01} = 6.2°$; time of flight to the target, $t_T = 17s$; terminal angle, $\alpha_T = 10°$; trajectory maximum altitude, $y_m = 345m$.

Find the departure angle needed to hit the target located at the horizontal range of 10,000 meters if there is a change of +10 degrees Celsius in the air temperature of air and at the same time a change of -10 mm Hg in the air pressure of air. Consider the air humidity 80%.

Solution:

Method 1

(a) First, we consider that the pressure remains unchanged, $p_2 = 750mm\ Hg$.

From table 2, section 2.2, we find that the pressure of water vapors is $e = 19mmHg$. The virtual temperature is

$$\tau_2 = \frac{T_2}{(1 - 0.3785 \cdot e_2 / p_2)} = \frac{298.15}{(1 - 0.3785 \cdot (19) / (750))} = 301.04°\ K$$

We have

$$J_2 = \frac{p_{0N}}{p_2}\sqrt{\frac{\tau_2}{\tau_{0N}}} = \frac{750}{750}\sqrt{\frac{301.04}{289.08}} = 1.02109,$$

$$X_{T2} = J_2 \cdot x_T = 1.02109 \cdot (10000) = 10210.9m,$$

and

$$\alpha_{02} = ar\sin(J \cdot \sin\alpha_{01}) = \arcsin(1.02047 \cdot \sin(6.2°)) = 6.33127° .$$

In a similar way, we find as well the data that corresponds to a change -10 degrees Celsius, i.e., $T_2 = 278.15°\ K$. We have:

$$\tau_2 = \frac{T_2}{(1 - 0.3785 \cdot e_2 / p_2)} = \frac{278.15}{(1 - 0.3785 \cdot (5.232) / (750))} = 278.89°\ K,$$

$$J_1 = \frac{p_{0N}}{p_2}\sqrt{\frac{\tau_2}{\tau_{0N}}} = \frac{750}{750}\sqrt{\frac{278.89}{289.08}} = 0.98221,$$

$$X_{T1} = J_1 \cdot x_T = 0.98221 \cdot (10000) = 9822.2m$$

and

$$\alpha_{02} = ar\sin(J \cdot \sin\alpha_{01}) = \arcsin(0.98221 \cdot \sin(6.2°)) = 6.08932° .$$

The "correction" in departure angle needed to hit the target at the distance of 10,000 meters is

$$\Delta \alpha_t = \frac{\alpha_2 - \alpha_1}{X_{T2} - X_{T1}} \cdot h = \frac{(6.33127° - 6.08932°)}{(10210.9 - 9822.11)} \cdot (10000 - 10210.9) = -0.13124° \cdot$$

(b) The air pressure changes from 750mm Hg to 740mm Hg, and the temperature remains constant, 15 degrees Celsius, while the humidity is 80%.

For $p_2 = 740mm\ Hg$, the virtual temperature is

$$\tau_2 = \frac{T_2}{(1 - 0.3785 \cdot e_2 / p_2)} = \frac{288.15}{(1 - 0.3785 \cdot (10.16) / 740)} = 289.66° \ K.$$

We have:

$$J_2 = \frac{p_{0N}}{p_2} \sqrt{\frac{\tau_2}{\tau_{0N}}} = \frac{750}{740} \sqrt{\frac{289.66}{289.08}} = 1.01453,$$

$$X_{T2} = J_2 \cdot x_T = 1.01453 \cdot (10000) = 10145.3m$$

and

$$\alpha_{02} = ar\sin(J \cdot \sin\alpha_{01}) = \arcsin(1.01452 \cdot \sin(6.2°)) = 6.29045°.$$

In the same way, we find the horizontal range and the departure angle when the pressure changes with +10mm Hg, i.e., when $p_2 = 760mm\ Hg$.

The virtual temperature is

$$\tau_2 = \frac{T_2}{(1 - 0.3785 \cdot e_2 / p_2)} = \frac{288.15}{(1 - 0.3785 \cdot (10.16) / 760)} = 289.62° \ K.$$

We have:

$$J_1 = \frac{p_{0N}}{p_2} \sqrt{\frac{\tau_2}{\tau_{0N}}} = \frac{750}{760} \sqrt{\frac{289.62}{289.08}} = 0.98776,$$

$$X_{T1} = J_1 \cdot x_T = 0.98776 \cdot (10000) = 9877.6m$$

and

$$\alpha_{02} = ar\sin(J \cdot \sin\alpha_{01}) = \arcsin(0.98776 \cdot \sin(6.2°)) = 6.12384°.$$

The "correction" in departure angle needed to hit the target at the distance of 10,000 meters is

$$\Delta\alpha_p = \frac{\alpha_2 - \alpha_1}{X_{T2} - X_{T1}} \cdot h = \frac{(6.29045° - 6.12384°)}{(10145.3 - 9877.6)} \cdot (10000 - 10145.3) = -0.09043° \cdot$$

Correction in departure angle

The total correction in departure angle is

$$\Delta\alpha_0 = \Delta\alpha_t + \Delta\alpha_p = -0.13124° + (-0.09043°) = -0.22167°.$$

The corrected departure angle is

$$\alpha_{02} = \alpha_0 + \Delta\alpha_0 = 6.2 + (-0.22167°) = 5.97833°.$$

Method 2

(a) Consider the changes in temperature and pressure are respectively +10 degree Celsius and -10mm Hg. The actual temperature and pressure are respectively $T_2 = 298.15°$ and $p_2 = 740mmHg$. The virtual temperature is

$$\tau_2 = \frac{T_2}{(1 - 0.3785 \cdot e_2 / p_2)} = \frac{298.15}{(1 - 0.3785 \cdot (19) / (740))} = 301.08° K.$$

We have:

$$J_2 = \frac{p_{0N}}{p_2}\sqrt{\frac{\tau_2}{\tau_{0N}}} = \frac{750}{740}\sqrt{\frac{301.08}{289.08}} = 1.03433,$$

$$X_{T2} = J_2 \cdot x_T = 1.03433 \cdot (10000) = 10343.3m,$$

$$\alpha_{02} = ar\sin(J \cdot \sin\alpha_{01}) = \arcsin(1.03433 \cdot \sin(6.2°)) = 6.41372°.$$

(b) Consider the changes in temperature and pressure respectively -10 degrees Celsius and -10mm Hg. The actual temperature and the actual pressure are respectively $T_2 = 278.15°$ and $p_2 = 760mm\ Hg$.

The virtual temperature is

$$\tau_2 = \frac{T_2}{(1 - 0.3785 \cdot e_2 / p_2)} = \frac{278.15}{(1 - 0.3785 \cdot (5.232) / (760))} = 278.88°\ K \cdot$$

We have:

$$J_1 = \frac{p_{0N}}{p_2} \sqrt{\frac{\tau_2}{\tau_{0N}}} = \frac{750}{760} \sqrt{\frac{278.88}{289.08}} = 0.99547 \,,$$

$$X_{T1} = J_1 \cdot x_T = 0.99547 \cdot (10000) = 9954.7m \,,$$

$$\alpha_{02} = ar\sin(J \cdot \sin\alpha_{01}) = arcsin(0.99547 \cdot \sin(6.2°\,)) = 6.17178°\,.$$

Correction in departure angle

The total "correction" in departure angle needed to hit the target at the distance of 10,000 meters is

$$\Delta\alpha_p = \frac{\alpha_2 - \alpha_1}{X_{T2} - X_{T1}} \cdot h = \frac{(6.41372° - 6.17178°)}{(10343.3 - 9954.7)} \cdot (10000 - 10343.3) = -0.21373° \cdot$$

The departure angle is

$$\alpha_{02} = \alpha_0 + \Delta\alpha_0 = 6.2 + (-0.21373°\,) = 5.9863°\,.$$

In the following table there are given the data obtained using the above method as well as the data calculated using the range table of 162mm cannon 162mm and the data calculated using the PC program Aca122.Bas.

Table 1: Range 10,000 meters

	Angle	Correction
Modern Method	5.9863	-0.21373
Range Table	6.0284	-0.17158
ACA122.Bas	6.0690	-0.13293

The results obtained using method 1 and method 2 are approximately the same.

4.2. Construction of Non-Standard Atmosphere Range Tables

One of the main problems of exterior ballistics is the calculation of the projectile trajectory in non standard atmosphere, including the calculation of the firing range in mountains.

In general, for each firearm and a particular projectile, there is given the range table in standard atmosphere that contains as well the range corrections for small deviations of the temperature, pressure, and humidity of air from the standard values is given. Based on those corrections, we are able to prepare the initial data for shooting in atmospheric conditions that are close to the sea -level standard conditions.

For shooting in high mountains, where the deviations in pressure and temperature are relatively large, the sea -level range tables are useless. That's why in artillery there are constructed as well the mountain range tables for altitudes 500m, 1,000m, 1,500m, 2,000m, 2,500m, 3,000m are constructed as well.

In this section, we show a new practical approach to prepare the firing data and the range tables for the sea-level non standard atmosphere and for the mountain atmosphere, based on the sea-level standard range tables.

The method presented hereafter is not only much simpler thatn the method of linear approximation introduced in the following sections, but can be used as well when the deviations of the characteristics of the atmosphere from the standard values are relatively large.

We assume that the ballistics characteristics of the projectile and the propellant are standard and that the weather is without wind.

As in section 4.1, we will use the similitude formulae:

Horizontal range:

$$X_T = J \cdot x_T, \tag{4.2.1}$$

Departure angle:

$$\sin \alpha_{02} = J \cdot \sin \alpha_0, \quad \alpha_{02} = \arcsin(J \cdot \sin \alpha_0), \tag{4.2.2}$$

Time of flight:

$$t_{T2} = J \cdot t_T, \tag{4.2.3}$$

Terminal angle:

$$\alpha_{T2} = ar \sin(J \cdot \sin \alpha_T). \tag{4.2.4}$$

Trajectory vertex:

$$Y_m = J^2 y_m, \tag{4.2.5}$$

where

$$J = \frac{p_{0N}}{p_2} \sqrt{\frac{\tau_2}{\tau_{0N}}}, \tag{4.2.6}$$

while p_{0N}, τ_{0N} are respectively the virtual temperature and pressure for the standard atmosphere, while p_2, τ_2 are the virtual temperature and pressure of air in non standard atmosphere.

Consider that we want to calculate the firing data in non-standard atmosphere to hit the target located at the horizontal range X_T. The corresponding similitude horizontal range in the standard atmosphere is

$$x_T = \frac{X_T}{J}.$$ (4.2.7)

At the standard range table, the above value is between the horizontal ranges x_{T1} and x_{T2}. Thus, the horizontal range X_T is located between

$$X_{T1} = J \cdot x_{T1} \text{ and } X_{T2} = J \cdot x_{T2}.$$ (4.2.8)

Using the similitude formulas and the standard range table, we are able to calculate the elements of the trajectory that correspond to the horizontal ranges given by (4.2.8). Then using the linear interpolation, we can find the elements of the trajectory that corresponds to the horizontal range X_T.

For example, to find the departure angle α_{02} that corresponds to X_T, we use the following formula:

$$\alpha_{02} = \alpha_1 + \frac{\alpha_2 - \alpha_1}{(X_{T2} - X_{T1})}(X_T - X_{T1}),$$ (4.2.9)

where α_1 and α_2 are the departure angles that correspond respectively to the horizontal ranges X_{T1} and X_{T2}. Angles α_1 and α_2 are computed using (4.2.2).

The above formula can be simplified and presented in the following form:

In a similar way we can compute the time of flight T_F, coordinates of vertex, X_m and Y_m, terminal angle, α_{T2}, and terminal speed V_2 using the following formulae:

Time of Flight:

$$T_F = J \cdot t_1 + J \frac{t_2 - t_1}{(x_{T2} - x_{T1})}(x_T - x_{T1}).$$ (4.2.10)

Coordinates of Vertex:

$$X_m = J \cdot x_{m1} + J \frac{x_{m2} - x_{m1}}{(x_{T2} - x_{T1})}(x_T - x_{T1}),$$ (4.2.11)

and

$$Y_m = J^2 y_{m1} + J^2 \frac{y_{m2} - y_{m1}}{(x_{T2} - x_{T1})}(x_T - x_{T1}).$$ (4.2.12)

Terminal Speed:

$$V_2 = v_1 + \frac{v_2 - v_1}{(x_{T2} - x_{T1})}(x_T - x_{T1}).$$ (4.2.12)

Terminal Angle:

$$\alpha_{T2} = \alpha_1 + \frac{\alpha_2 - \alpha_1}{(x_{T2} - x_{T1})}(x_T - x_{T1}),$$ (4.2.13)

where, in this case, α_1 and α_2 are the terminal angles that correspond respectively to the horizontal ranges X_{T1} and X_{T2}.

The correction method presented in the above two sections calculates not only the range corrections but as well the corrections for other elements of the trajectory as well, which is unusual for the existing range tables.

Example2.1 Projectile Trajectory in Non-standard Atmosphere

For the 122mm Russian projectile, fired by a Model 1960 cannon, Model 1960 at a speed of 885m/s, find the departure angle α_2 needed to hit the target located at the horizontal range of 10,000 meters as well as the terminal speed V_2 and the maximum altitude Y_m if the atmospheric pressure is 740 mm Hg, the temperature of air is +25 degrees Celsius, and the humidity is 80%.

Consider a non-windy weather and that the ballistics characteristics of the projectile and the propellant charge are standard.

Solution:

Preparatory Data

The standard atmosphere for the sea-level range table is TSA: virtual temperature $\tau_{0N} = 289.08° K$, temperature of dry air is 15 degrees Celsius, relative humidity of air is 50%, and pressure of water vapors, is 6.35mm Hg.

The value of the parameter J, calculated in example 1.3 is:

$$J = \frac{p_{0N}}{p_2} \sqrt{\frac{\tau_2}{\tau_{0N}}} = \frac{750}{740} \sqrt{\frac{301.08}{289.08}} = 1.03433 \,.$$

Using (4.2.7), we find that

$$x_T = \frac{X_T}{J} = \frac{10000}{1.0343} = 9668.09 \,.$$

Thus, the above value is between $x_{T1} = 9600$ and $x_{T1} = 9800$.

From the sea-level range table of the given cannon, we find the following data:

(a) For the horizontal range $x_{T1} = 9600m$, we have: $\alpha_{01} = 5.8°$, $v_1 = 429m/s$

(b) and $y_m = 305m$.

(c) For the horizontal range $x_{T2} = 9800m$, we have: $\alpha_{02} = 6°.00$, $v_2 = 422m/s$, and $y_{m2} = 325m$

The Departure Angle

To find that the required departure angles, we use the formula (4.2.9).

We find that the similitude departure angles that correspond to the departure angles given in (a) and (b) are respectively:

$$\alpha_1 = \arcsin(J \cdot \sin\alpha_{01}) = \arcsin[(1.0343) \cdot \sin(5.8°)] = 5.9998°$$

and
$$\alpha_2 = \arcsin(J \cdot \sin \alpha_{02}) = \arcsin[(1.0343) \cdot \sin(6.00°)] = 6.2068° \, .$$

The departure angle needed to hit the target, in the given non-standard atmosphere, at the horizontal range of 10,000 meters, is:

$$\alpha_{02} = \alpha_1 + \frac{\alpha_2 - \alpha_1}{(x_{T2} - x_{T1})}(x_T - x_{T1}) =$$

$$= 5.9998 + \frac{6.2068° - 5.9998°}{9800 - 9600} \cdot (9668.09 - 9600) = 6.0703° \, .$$

The Terminal Speed

$$V_2 = v_1 + \frac{v_2 - v_1}{(x_{T2} - x_{T1})}(x_T - x_{T1}) =$$

$$= 429 + \frac{422 - 429}{9800 - 9600}(9668.09 - 9600) = 426.62 m / s.$$

The Maximum Altitude

$$Y_{m2} = J^2 y_{m1} + J^2 \frac{y_{m2} - y_{m1}}{(x_{T2} - x_{T1})}(x_T - x_{T1}) =$$

$$= (1.03433)^2 \cdot (305) + (1.03433)^2 \cdot \frac{325 - 305}{9800 - 9600} \cdot (9668.09 - 9600) = 312.28m$$

Using the range table of the Russian 122mm cannon, we find that the departure angle is $\alpha_{02} = 6.03°$.

Example 2.2 Construction of Mountain Range Tables

A 122mm projectile of a Model 1960 Russian cannon Model 1960 is to be fired at a speed of 885m/s, at the altitude of 1,500 meters over the sea level, in order to hit a target located at the horizontal distance of 14,400 meters. The temperature, pressure, and the humidity of air at the firing site are respectively 6 degrees, 626 mm Hg, and 50%.

Use the sea-level range table of the given cannon to find the departure angle α_{02}, the time of flight T_F, coordinates of vertex,

X_m and Y_m, terminal angle, α_{T2}, and terminal speed V_T that correspond to the mountain horizontal range $X_T = 14400m$.

Preparatory Data

The standard atmosphere at the sea -level range table is TSA: virtual temperature is $\tau_{0N} = 289.08° K$, temperature of dry air is 15 degrees Celsius, relative humidity of air is 50%, and pressure of water vapors, is 6.35mm Hg.

The actual temperature and pressure at the firing site are respectively $T_2 = 279.15° K$ and $p_2 = 626mm$ Hg.

The virtual temperature is

$$\tau_2 = \frac{T_2}{(1 - 0.3785 \cdot e_2 / p_2)} = \frac{279.15}{(1 - 0.3785 \cdot (3.536) / (626))} = 279.75° K.$$

We have:

$$J = \frac{p_{0N}}{p_2}\sqrt{\frac{\tau_2}{\tau_{0N}}} = \frac{750}{626}\sqrt{\frac{279.75}{289.08}} = 1.17859.$$

Using (4.2.7), we find that

$$x_T = \frac{X_T}{J} = \frac{14400}{1.17859} = 12218m.$$

Thus, the above value is between $x_{T1} = 12200m$ and $x_{T1} = 12400$. From the sea -level range table of the given cannon, we find the following data:

(a) For the horizontal range $x_{T1} = 12200m$: $\alpha_{01} = 8.8833°$,

    $x_{m1} = 7040m$, $y_{m1} = 649m$, $t_1 = 23$, $v_1 = 351m/s$,

    $\alpha_{T1} = 17.00°$.

(b) For the horizontal range $x_{T2} = 12400m$: $\alpha_{02} = 9.1667°$,

    $x_{m2} = 7040m$, $y_{m2} = 686m$, $t_2 = 24s$, $v_2 = 347m/s$,

    $\alpha_{T2} = 17.00°$.

The Departure Angle

We find that the similitude departure angles that correspond to the departure angles given in (a) and (b) are respectively:

$$\alpha_1 = \arcsin(J \cdot \sin \alpha_{01}) = \arcsin[(1.17859) \cdot \sin(8.8833°)] = 10.48632°$$

$$\alpha_2 = \arcsin(J \cdot \sin \alpha_{02}) = \arcsin[(1.17859) \cdot \sin(9.1667°)] = 10.8212° .$$

The departure angle needed to hit the target at the horizontal range of 14,400 meters in the given non-standard atmosphere is:

$$\alpha_{02} = \alpha_1 + \frac{\alpha_2 - \alpha_1}{(x_{T2} - x_{T1})}(x_T - x_{T1}) =$$

$$= 10.48632 + \frac{10.8212° - 10.4863°}{12400 - 12200} \cdot (12218 - 12200) = 10.5165°$$

The Terminal Speed

$$V_2 = v_1 + \frac{v_2 - v_1}{(x_{T2} - x_{T1})}(x_T - x_{T1}) =$$

$$= 351 + \frac{347 - 351}{12400 - 12200}(12218 - 12000) = 350.64 m/s$$

The Maximum Altitude

$$Y_m = J^2 y_{m1} + J^2 \frac{y_{m2} - y_{m1}}{(x_{T2} - x_{T1})}(x_T - x_{T1}) = (1.17859)^2 \cdot (649) +$$

$$+ (1.17859)^2 \cdot \frac{686 - 649}{12400 - 12000} \cdot (12218 - 12000) = 906.51 m$$

The Time of Flight

$$T_F = J \cdot t_1 + J \frac{t_2 - t_1}{(x_{T2} - x_{T1})}(x_T - x_{T1}) =$$

$$= (1.17859)[(23) + \frac{24 - 23}{12400 - 12200}(12218 - 12200)] = 27.21$$

Note that the terminal angle remains practically the same.

In the following table, are shown the data obtained above as well as the data obtained by the mountain table of the 122mm Russian cannon Model 1960, and the data obtained using Aka122.Bas are shown.

Table 1: Range 14,400 meters, altitude of the firing site 1,500 meters.

	Angle	Speed	Altitude	Time
New Method	10.5165	351	907	27.21
Range Table	10.5667	353	916	27
ACA122.Bas	10.5686	364	905	26.95
Ballistica22	10.5842	352		2.71

Example 2.3 Accuracy of Snell's Law

In table 1 it is displayed again the range table of the bullet 0.338 Lapua GB528 Scenar 19.44g (300 grain) fired with a muzzle speed of 830m/s, at the sea level in the ICAO atmosphere (pressure, 760 mm Hg,; temperature, 15 degrees Celsius,; density of air, 1.225kg/m^3,; speed of sound, 340.30m/s,; relative humidity, 0%).
(Reference: Example 8.1, section 2.8).
Table 1 is obtained using the PC program "Ballistica2.2" (English version 2.2) of Jan Krčmář, Ph.D. [69]
The bullet has a mass of 19.44 grams, and a diameter of 8.6mm. The form coefficient with respect to the drag function is 1, and the ballistics coefficient in metric units is 3.8037.

Table 1: Range Table, Bullet 0.338 Lapua GB528 Scenar 19.44g; sea level

Range [m]	100	200	300	400	500	600
Initial Angle	0.0422	0.0873	0.1357	0.1877	0.2436	0.3037
Impact Speed	789	750	712	675	639	605
Time	0.124	0.254	0.392	0.536	0.688	0.849

---

[69]. Jan Krčmář , Ballistica2.2 http://www.balistika.cz/eng/exterior.html (PC program downloaded on 6 November, 2009 and used with author's permission)

Table 1: continue

Range [m]	700	800	900	1000	1100	1200
Initial Angle	0.3687	0.4389	0.5151	0.5977	0.6878	0.786
Impact Speed	572	539	508	478	449	422
Time	1.019	1.200	1.391	1.593	1.809	2.038

Use the data of table 1 to find the elements of the projectile trajectory for the horizontal range of 1,000 meters if the firing site is located at an altitude of 1,800 meters. The temperature, the pressure, and the humidity of air are respectively -7.4 degrees Celsius, 603mm Hg, and 50%.

Solution:

Preparatory Data

The pressure at the firing site is $p_2 = 603mmHg$. The virtual temperature at the firing site is

$$\tau_2 = \frac{T_2}{(1 - 0.3785 \cdot e_2 / p_2)} = \frac{273.15 + (-7.4)}{(1 - 0.3785 \cdot (1.32) / (603))} = 266° K.$$

The virtual temperature and pressure at the sea level (ICAO atmosphere) are respectively

$$\tau_{0N} = 273.15 + 15 = 288.15° K \text{ and } p_{0N} = 760mmHg.$$

We have:

$$J = \frac{p_{0N}}{p_2}\sqrt{\frac{\tau_2}{\tau_{0N}}} = \frac{760}{603}\sqrt{\frac{266}{288.15}} = 1.2109.$$

Using (4.2.7), we find that

$$x_T = \frac{X_T}{J} = \frac{1000}{1.2109} = 825.84m.$$

Thus, the above value is between $x_{T1} = 800m$ and $x_{T1} = 900m$.

From table 1, we have:

(a) for the horizontal range $x_{T1} = 800m : \alpha_{01} = 0.4389°$, $t_1 = 1.2s$,
$v_1 = 539m/s$.

(b) for the horizontal range $x_{T2} = 900m : \alpha_{02} = 0.5151°$,
$t_2 = 1.39s$, $v_2 = 508m/s$.

The Departure Angle

We find that the similitude departure angles that correspond to the departure angles given in (a) and (b) are respectively:

$$\alpha_1 = \arcsin(J \cdot \sin\alpha_{01}) = \arcsin[(1.2109) \cdot \sin(0.4389°)] = 0.5315°$$

$$\alpha_2 = \arcsin(J \cdot \sin\alpha_{02}) = \arcsin[(1.2109) \cdot \sin(0.5151°)] = 0.6237° .$$

The departure angle needed to hit the target at the horizontal range of 1,000 meters (altitude of 1,800 meters over the sea level) is:

$$\alpha_0 = \alpha_1 + \frac{\alpha_2 - \alpha_1}{(x_{T2} - x_{T1})}(x_T - x_{T1}) =$$

$$= 0.5315 + \frac{0.6237° - 0.5315°}{900 - 800} \cdot (825.84 - 800) = 0.5553°$$

The Terminal Speed

$$V_2 = v_1 + \frac{v_2 - v_1}{(x_{T2} - x_{T1})}(x_T - x_{T1}) =$$

$$= 539 + \frac{508 - 539}{900 - 800}(825.84 - 800) = 531m/s$$

The Time of Flight

$$T_F = J \cdot t_1 + J\frac{t_2 - t_1}{(x_{T2} - x_{T1})}(x_T - x_{T1}) =$$

$$= (1.2109)[(1.2) + \frac{1.39 - 1.2}{900 - 800}(825.84 - 800)] = 1.513s$$

In the following table are shown the data obtained above as well as the data obtained using the PC program "Ballistica2.2" (English version 2.2) of Jan Krčmář [70].

Table 1: Range 1,000 meters, altitude 1,800m

	Angle	Speed	Time y_m
Modern Method	0.5553	531	1.513
Ballistica2.2	0.5601	524	1.522

Comment:

The results obtained using the new approach based on the similitude laws and Snell's law, confirm once again the accuracy of Snell's law approach to find the departure angle.

4.3. Converting Firing Data to Standard Atmosphere Values

One of the main tasks of the exterior ballistics is the construction of the sea-level standard range tables using a theoretical approach to manipulate the experimental data that are not obtained in standard atmospheric conditions.

The standard atmospheric conditions are not always present at the ballistics proving ground. Since the shooting tests are usually carried out in non standard atmospheric conditions, the horizontal range and all the other elements of the trajectory, obtained in the actual non standard atmosphere need to be "converted", or to be "brought" in standard atmospheric conditions.

In other words, using the experimental data for the elements of the trajectory, we have to determine what would have been the horizontal range and other elements of the trajectory if the firearm

---

70.   Jan Krčmář , Ballistica2.2 http://www.balistika.cz/eng/exterior.html (PC program downloaded on 6 November, 2009 and used with author's permission).

tests were performed in standard atmospheric conditions (wind is not present).

In "*Exterior Ballistics with Applications*, section 6.2," we used the laborious correction theory based on the approximate Siacci's method to "bring" the experimental data into standard atmosphere.

The simple and practical new method introduced in section 4.2 can be used to convert the experimental firing data into standard values and to construct the sea -level standard atmosphere range tables.

It can also be used to convert the data of experimental shooting, performed in high -altitude sites and non-standard atmosphere, into standard firing data appropriate for the standard atmosphere at the firing site.

A mountain or high -altitude atmosphere is called standard atmosphere if the temperature and pressure at the given altitude match the respective data that can be obtained using the accepted laws of change of the temperature, density, and pressure with altitude y over the sea level:

$$\tau = \tau_0 - 0.006328y, \tag{4.3.1}$$

$$\rho = \rho_0 \left(\frac{\tau_0 - 0.006328y}{\tau_0}\right)^{4.4}, \tag{4.3.2}$$

$$p = p_0 \left(\frac{\tau_0 - 0.006328y}{\tau_0}\right)^{5.4}, \tag{4.3.3}$$

when the temperature, the density, and pressure at the sea level are standard. Note that we find the virtual temperature using table 1 section 2.2.

For approximate calculations, the air pressure at the altitude y over the sea level can be determined approximately using (2.2.8),

$$p = p_0 - 8.5 \cdot (y - y_0) / 100. \tag{4.3.4}$$

For example, for the ICAO standard atmosphere the temperature, the pressure, the density and the humidity at the sea level are respectively:

$t_0 = 15°C$, $p_0 = 760mm$ $Hg.$, $\rho_{0N} = 1.225 kg / m^3$, and 0%.

At the altitude $y = 2000m$, the ICAO standard values of the temperature, density, and the air pressure (calculated using the above formulae) are respectively $t_0 = 2.344° C$,

$p_0 = 593.65 mmHg$, and $\rho_0 = 1.005 kg / m^3$.

We illustrate the method presented above through the following examples.

Example 3.1 Conversion of the Experimental Data to Mountain Values

A ballistician has to compile a mountain standard range table for the bullet 0.338 Lapua GB528 Scenar 19.44g (300 grain) for the altitude of 500 meters over the sea level. The proving ground is at the altitude 320 meters over the sea level.

Assume that the measured temperature, pressure, and humidity (at the proving ground at the time of firing tests) are respectively 9.92 degrees Celsius, 725.53 mm Hg, and 60%. There is no wind, and the temperature of the projectile (propellant charge) is kept at 15 degrees Celsius. The projectile initial speed is 830m/s.

The standard ICAO values of the temperature, pressure, and humidity at the altitude of 500 meters are respectively 11.75 degrees Celsius, 716 mm Hg, and 0% (assuming that at the sea level, the values of the temperature, the pressure and humidity are respectively 15 degrees Celsius, 760 mm Hg, and 0%).

The ballistician wants to find the elements of the trajectory at the horizontal range of 500 meters at the altitude of 500 meters using firing tests at 320 meters.

Estimating the parameter J, he decides to do the firing tests for horizontal ranges od 450m and 550m.

The results of the firing tests are:

(a) For the horizontal range $x_{T1} = 450m : \alpha_{01} = 0.214°$,

$x_{m1} = 235m, y_{m1} = 0.47m, t_1 = 0.608s, v_1 = 662m/s$.

(b) For the horizontal range $x_{T2} = 550m : \alpha_{02} = 0.2713°$,

$x_{m2} = 282m, y_{m2} = 0.72m, t_2 = 0.764s, v_2 = 628m/s$.

What are the elements of the trajectory for the horizontal range of 500 meters at the altitude of 500 meters in the standard ICAO atmosphere?

Solution:

Preparatory Data

The virtual temperature at the firing site is

$$\tau_1 = \frac{T_1}{(1 - 0.3785 \cdot e_1 / p_1)} = \frac{273.15 + 9.92}{(1 - 0.3785 \cdot (5.5)/(716))} = 283.9° \ K \ .$$

The virtual temperature at the altitude of 500 meters is equal to the temperature

$$\tau_2 = 273.15 + 11.75 = 284.9° \ K \ .$$

We have:

$$J = \frac{p_1}{p_2} \sqrt{\frac{\tau_2}{\tau_1}} = \frac{725.53}{716} \sqrt{\frac{284.9}{283.9}} = 1.0151 \ .$$

Using (4.2.7), we find that

$$x_T = \frac{X_T}{J} = \frac{500}{1.17859} = 492.57m \ .$$

Thus, the above value is between $x_{T1} = 450m$ and $x_{T1} = 550$.

The Departure Angle

To find the required departure angles, we use the formula (4.2.9). We find that the similitude departure angles that correspond to the departure angles given in (a) and (b) are respectively:

$$\alpha_1 = \arcsin(J \cdot \sin\alpha_{01}) = \arcsin[(1.0151) \cdot \sin(0.214°)] = 0.2172°$$

$$\alpha_2 = \arcsin(J \cdot \sin\alpha_{02}) = \arcsin[(1.0151) \cdot \sin(0.2713°)] = 0.2754°.$$

The departure angle needed to hit the target at the horizontal range of 500 meters (altitude of 500 meters over the sea level, standard ICAO atmosphere) is:

$$\alpha_{02} = \alpha_1 + \frac{\alpha_2 - \alpha_1}{(x_{T2} - x_{T1})}(x_T - x_{T1}) =$$

$$= 0.2172 + \frac{0.2754° - 0.2172°}{550 - 450} \cdot (492.27 - 450) = 0.2418°$$

The Terminal Speed

$$V_2 = v_1 + \frac{v_2 - v_1}{(x_{T2} - x_{T1})}(x_T - x_{T1}) =$$

$$= 662 + \frac{628 - 662}{550 - 450}(492.27 - 450) = 647.63m/s$$

The Time of Flight

$$T_F = J \cdot t_1 + J\frac{t_2 - t_1}{(x_{T2} - x_{T1})}(x_T - x_{T1}) =$$

$$= (1.0151)[(0.608) + \frac{0.764 - 0.608}{550 - 450}(492.57 - 450)] = 0.685$$

The Coordinates of Vertex

$$X_m = J \cdot x_{m1} + J\frac{x_{m2} - x_{m1}}{(x_{T2} - x_{T1})}(x_T - x_{T1}) =$$

$$= (1.0151) \cdot [235 + \frac{282 - 235}{550 - 450} \cdot (492.57 - 450)] = 259m$$

and

$$Y_m = J^2 \cdot y_{m1} + J^2 \frac{y_{m2} - y_{m1}}{(x_{T2} - x_{T1})}(x_T - x_{T1}) =$$

$$= (1.0151)^2 \cdot [(0.47) + \frac{0.72 - 0.47}{550 - 450} \cdot (492.57 - 450) = 0.576m$$

In the following table, are shown the data obtained above as well as the data obtained using the PC program "Ballistica2.2" (English version 2.2) of Jan Krčmář [71] are shown.

Table 1: Range, 500 meters,; altitude of the firing site, 500 meters

	Angle	Speed	Time	x_m	y_m
Modern Method	0.2418	647.6	0.685	259	0.576
Ballistica2.2	0.2421	645	0.684	258	0.58

Example 3.2 Converting Experimental Data to Sea level Values

Use the data of example 3.1 to find the elements of the projectile for the horizontal range of 500 meters at the sea level in ICAO atmosphere.

Solution:

Using the data given in Example 3.1, we find:

$$J = \frac{p_1}{p_2}\sqrt{\frac{\tau_2}{\tau_1}} = \frac{725.53}{760}\sqrt{\frac{288.15}{283.9}} = 0.96176.$$

Using (4.2.7), we find that

$$x_T = \frac{X_T}{J} = \frac{500}{0.96176} = 519.88m.$$

---

71. Jan Krčmář, Ballistica2.2 http://www.balistika.cz/eng/exterior.html (PC program downloaded on 6 November, 2009 and used with author's permission).

The above value is between $x_{T1} = 450m$ and $x_{T1} = 550$.

The Departure Angle

We find that the similitude departure angles that correspond to the departure angles given in (a) and (b) of Example 3.1 are respectively:

$$\alpha_1 = \arcsin(J \cdot \sin\alpha_{01}) = \arcsin[(0.96176) \cdot \sin(0.214°)] = 0.2058°$$

and

$$\alpha_2 = \arcsin(J \cdot \sin\alpha_{02}) = \arcsin[(0.96176) \cdot \sin(0.2713°)] = 0.2609°.$$

The departure angle needed to hit the target at the horizontal range of 500 meters at the sea level standard atmosphere is:

$$\alpha_{02} = \alpha_1 + \frac{\alpha_2 - \alpha_1}{(x_{T2} - x_{T1})}(x_T - x_{T1}) =$$

$$= 0.2058 + \frac{0.2609° - 0.2058°}{550 - 450} \cdot (519.88 - 450) = 0.2443°.$$

The Terminal Speed

$$V_2 = v_1 + \frac{v_2 - v_1}{(x_{T2} - x_{T1})}(x_T - x_{T1}) =$$

$$= 662 + \frac{628 - 662}{550 - 450}(519.88 - 450) = 638.24m/s.$$

The Time of Flight

$$T_F = J \cdot t_1 + J\frac{t_2 - t_1}{(x_{T2} - x_{T1})}(x_T - x_{T1}) =$$

$$= (0.9618)[(0.608) + \frac{0.764 - 0.608}{550 - 450}(519.88 - 450)] = 0.69.$$

The Coordinates of Vertex

$$X_m = J \cdot x_{m1} + J \frac{x_{m2} - x_{m1}}{(x_{T2} - x_{T1})}(x_T - x_{T1}) =$$

$$= (0.9618) \cdot [235 + \frac{282 - 235}{550 - 450} \cdot (519.88 - 450)] = 257.6m$$

and

$$Y_m = J^2 \cdot y_{m1} + J^2 \frac{y_{m2} - y_{m1}}{(x_{T2} - x_{T1})}(x_T - x_{T1}) =$$

$$= (0.9618)^2 \cdot [(0.47) + \frac{0.72 - 0.47}{550 - 450} \cdot (519.88 - 450) = 0.60m.$$

In the following table are shown the data obtained above as well as the data obtained using the PC program "Ballistica2.2" (English version 2.2) of Jan Krčmář.

Table 1: Range, 500 meters,; altitude of the firing site, 500 meters

	Angle	Speed	Time	x_m	y_m
Modern Method	0.2443	638	0.69	258	0.60
Ballistica2.2	0.2437	639	0.688	258	0.58

4.4. Siacci's Method for Range Corrections

The traditional theory of corrections is based on the use of the ballistics tables, or in the Siacci's solution method of differential equations of projectile flight. [72]. Some elements of the theory of corrections are presented in chapter 6 of the "*Exterior Ballistics with Applications*".

In this section, we update and enrich the traditional method of Siacci's theory of corrections that is presented in "*Exterior Ballistics with Applications*".

The trajectory of flight of a projectile in a given standard atmosphere and for standard ballistics characteristics of the projectile is determined by the departure speed v_0, the departure angle α_0, and

---

72. Shapiro, J., "*Exterior Ballistics*", p. 174, Oborongiz, Moscow, '50.

the ballistics coefficient c. Thus, the horizontal range x_T of a projectile in standard atmosphere is determined by v_0, α_0, and c.

In general, for the flight of a projectile in a standard atmosphere, the (standard) horizontal range is a function of the initial speed, launching angle, ballistics coefficient, air temperature τ_0, atmospheric pressure p_0, temperature of the projectile propellant T_c, etc., i.e.,

$$x_T = f(v_0, \alpha_0, c, \tau_0, p_0, T_c, ...). \qquad (4.4.1)$$

Traditionally, for practical purposes, exterior ballistics prepares range tables for a standard projectile fired in a standard atmosphere (ICAO, ASM, or TSA).

Those tables contain as well the range corrections for small deviations of the ballistics characteristics of projectile and atmosphere from the standard values. [73].

Horizontal Range Correction

Consider the standard trajectory of a projectile and the horizontal range (4.4.1). Consider as well some relatively small changes Δv_0, $\Delta \alpha_0$, Δc, Δm, $\Delta \tau_0$, Δp_0, ΔT_c, etc., respectively in initial speed "v_0", departure angle "α_0", ballistics coefficient "c", projectile mass "m", virtual temperature "τ_0", atmospheric pressure p_0, and the temperature of projectile propellant T_c, etc.

The total change Δx_T in the horizontal range can be considered approximately equal to the total differential of the horizontal range "x_T", i.e.,

$$\Delta x_T = \frac{\partial x_T}{\partial v_0} \Delta v_0 + \frac{\partial x_T}{\partial \alpha_0} \Delta \alpha_0 + \frac{\partial x_T}{\partial c} \Delta c + \frac{\partial x_T}{\partial \tau_0} \Delta \tau_0 + \frac{\partial x_T}{\partial p_0} \Delta p_0 + \frac{\partial x_T}{\partial T_c} \Delta T_c + ...$$

$$(4.4.2)$$

---

73. Note that the standard temperature of the propellant charge is 70 degrees Fahrenheit (21.1 degrees Celsius).
   In ATS atmosphere, the temperature of the propellant charge of 15 degrees Celsius is considered standard.

(Note that in (3.4.2), changes in range that correspond to higher degree of magnitude, individual small changes are neglected).

The partial derivatives

$$\frac{\partial x_T}{\partial v_0}, \frac{\partial x_T}{\partial \alpha_0}, \frac{\partial x_T}{\partial c}, \frac{\partial x_T}{\partial T_0}, \frac{\partial x_T}{\partial p_0}, \frac{\partial x_T}{\partial T_c}, \dots \qquad (4.4.3)$$

are called "correction coefficients".

Each correction coefficients is numerically equal to the correction in horizontal range that is related to a unit change in the given parameter.

For example, a change of one degree Kelvin in the air temperature, $\Delta \tau_0 = 1$, corresponds to a change in the horizontal range of

$$\Delta x_T = \frac{\partial x_T}{\partial \tau_0} \Delta \tau_0 = \frac{\partial x_T}{\partial \tau_0}(1) = \frac{\partial x_T}{\partial \tau_0}$$

The correction coefficients,

$$\frac{\partial x_T}{\partial v_0}, \frac{\partial x_T}{\partial \alpha_0}, \frac{\partial x_T}{\partial c} \qquad (4.4.4)$$

that correspond respectively to the departure speed, departure angle, and ballistics coefficient, are called the main correction coefficients.

Recall that, the correction theory is valid for small changes that are within the following limits [74]:

$$\Delta v_0 = \pm 25 m/s, \ \Delta \alpha_0 = \pm 1.685°, \ \Delta c = \pm 10\% \cdot c$$

The main correction coefficients can be calculated using different methods. Hereafter, we use the Siacci's method.

For departure angles till around 10 degrees, the correction results that can be obtained using the relations presented in this section are accurately enough. [75].

[74.] Field Artillery, Volume 6, p. 134, DND Canada, 1992
[75.] Shapiro, J. M. "Exterior Ballistics", p. 190, Oborongiz, Moscow, 50'

Note that the traditional theory of corrections, presented in sections 4.4 and 4.5, can not be used to find the mountain shooting corrections based on the standard sea-level range tables. Nevertheless, the traditional correction theory can be used to calculate mountain firing corrections if it is known the standard mountain range table of the respective firearm is known.

Main Correction Coefficients

The main correction coefficient for the G-function,

$$G_D(v) = \begin{cases} A \cdot v^2 & \text{for} \quad v \le 256 m/s \\ E \cdot v - F & \text{for} \quad v > 256 m/s \end{cases}, \tag{4.4.5}$$

for departure angles till 15 degrees, can be calculated approximately using the following relationships[76]:

- The correction coefficient that corresponds to the initial speed is

$$\frac{\partial x_T}{\partial v_0} = \frac{v_0}{\beta \cdot c \cdot G_D(v_0)}(1 + \frac{\tan \alpha_0}{\tan|\alpha_T|} - \frac{g x_T}{v_0^2 \cos^2 \alpha_0 \tan \alpha_T}). \tag{4.4.6}$$

- The correction coefficient that corresponds to the departure angle is

$$\frac{\partial x_T}{\partial \alpha_0} = \frac{x_T \cos(2\alpha_0)}{\cos^2 \alpha_0 \tan|\alpha_T|}. \tag{4.4.7}$$

Note that the units of the above coefficient are (meter/radian). To transform the above correction coefficient from "meter/radian" units into "meter/degree" units, we multiply by the factor $(\pi / 180)$ the result obtained using (4.4.7).

- The correction coefficient that corresponds to the ballistics coefficient is

---

[76]   Shapiro, J. M.., Exterior Ballistics", p.188, Oborongiz, Moscow 50'

$$\frac{\partial x_T}{\partial c} = -(1 - \frac{\tan\alpha_0}{\tan|\alpha_T|})\frac{x_T}{c}. \qquad (4.4.8)$$

where

$$\beta = \frac{h(\bar{y})}{\sqrt{\cos\alpha_0}}, \quad \text{or} \quad \beta = h(\bar{y})/\cos\alpha_0, \qquad (4.4.9)$$

respectively for the horizontal fire (field artillery fire) and inclined fire (anti-aircraft fire);

$$h(\bar{y}) = (\frac{\tau_{0N} - 0.006328\bar{y}}{\tau_{0N}})^{4.4},$$

- $\bar{y} = (2/3)y_m$ for field artillery, target on the ground,
- $\bar{y} = (1/2)y_m$ for anti-aircraft artillery (uphill fire); target is on the ascending point of the trajectory,
- y_m is the maximum altitude of the projectile trajectory,
- τ_{0N} is the standard virtual temperature at the firing site.

Secondary Correction Coefficients

Correction Coefficients Related with Variation in Temperature and Pressure

- The correction coefficient that corresponds to the temperature of air

$$\frac{\partial x_T}{\partial \tau_0} = (1 - \frac{v_0}{2x}\frac{\partial x_T}{\partial v_0})\frac{x_T}{\tau_0}. \qquad (4.4.10)$$

- The correction coefficient that corresponds to the atmospheric pressure

$$\frac{\partial x_T}{\partial p_0} = -(1 - \frac{\tan\alpha_0}{\tan|\alpha_T|})\frac{x_T}{p_0}. \qquad (4.4.11)$$

**Correction Coefficient Related with Variation
in Projectile Mass**

- The correction coefficient that corresponds to the projectile
  mass

$$\frac{\partial x_T}{\partial m} = (1 - \frac{\tan\alpha_0}{\tan|\alpha_T|} - 0.4\cdot\frac{v_0}{x_T}\frac{\partial x_T}{\partial v_0})\frac{x_T}{m}. \qquad (4.4.12)$$

**Correction Coefficients Related with Variation in Propellant
Charge Characteristics** [77]

- The correction coefficient that corresponds to the
  temperature (T_c) of the propellant charge is

$$\frac{\partial x_T}{\partial T_c} = 0.001\cdot v_0\frac{\partial x_T}{\partial v_0}. \qquad (4.4.13)$$

For some type of propellants, there should be used other
coefficients that are slightly different from 0.001 should be used.

For example, for the 122m projectile of the Russian Model 1960
cannon, Model 1960 we use the coefficient 0.001285 instead of
0.001.

In this case (4.4.13), the correction coefficient is

$$\frac{\partial x_T}{\partial T_c} = 0.001285\cdot v_0\frac{\partial x_T}{\partial v_0}. \qquad (4.4.13a)$$

For some rifle powders, we can use the following relation
obtained using the data given in Ballistica22 [78]

---

77. The correction coefficients related with the ballistics characteristics of the
propellant charge are based on the data presented in the Field artillery, V.6,
Ballistics and Ammunition; DND Canada 1992, http://www.scribd.com/
doc/4934783/BALLISTICS-AND-AMMUNITION, (Web access December
20, 2009).
78. Jan Krčmář, Ballistica2.2, Powder Temperature and Barrel length,
http://www.balistika.cz/eng/exterior.html (accessed on 6 November, 2009)

$$\frac{\partial x_T}{\partial T_c} = 0.0014 \cdot v_0 \frac{\partial x_T}{\partial v_0}. \qquad (4.4.13b)$$

For IMR powders (*IMR* means Improved Military Rifles), using the data provided by Rinker, [79], we conclude that the correction coefficient is

$$\frac{\partial x_T}{\partial T_c} = 0.001766 \cdot v_0 \frac{\partial x_T}{\partial v_0}. \qquad (4.4.13c)$$

- The correction coefficient that corresponds to the mass (m_p) of the propellant charge is

$$\frac{\partial x_T}{\partial m_p} = 0.60 \cdot v_0 \frac{\partial x_T}{\partial v_0}, \qquad \frac{\partial x_T}{\partial m_p} = 0.70 \cdot v_0 \frac{\partial x_T}{\partial v_0}, \qquad (4.4.14)$$

respectively for "chord or tube" propellant charges and "slotted tube or multi-tube" charges.

- The correction coefficient that corresponds to the propellant size (S_p) is

$$\frac{\partial x_T}{\partial S_p} = -0.15 \cdot v_0 \frac{\partial x_T}{\partial v_0}, \qquad \frac{\partial x_T}{\partial S_p} = -0.30 \cdot v_0 \frac{\partial x_T}{\partial v_0}, \qquad (4.4.15)$$

respectively for "chord or tube" charges and "slotted tube or multi-tube" charges.

- The correction coefficient that corresponds to the chamber capacity C_C (propellant type: cord, tube, slotted tube, multi-tube) is

$$\frac{\partial x_T}{\partial C_C} = -0.25 \cdot v_0 \frac{\partial x_T}{\partial v_0}. \qquad (4.4.16)$$

---

79. Rinker, R.A. "Understanding Firearm Ballistics", p.162, 6th Edition, Mulberry House Publishing, 2005.

- Shot travel (S_H) correction coefficient is

$$\frac{\partial x_T}{\partial S_H} = 0.20 \cdot v_0 \frac{\partial x_T}{\partial v_0}. \tag{4.4.17}$$

Correction coefficient related with barrel length [80]

The length of the firearm barrel influences the departure speed of the projectile. The correction coefficient, related with a change in barrel length (from the standard length), depends on the rifle and the powder type.

For example, for the 308 Winchester (initial speed 834m/s, bullet mass, 0.168kg, powder mass, 0.045kg, Varget) , the correction coefficient related with a change in barrel length (from the standard length) is

$$\frac{\partial x_T}{\partial L_B} = 0.2031 \cdot v_0 \frac{\partial x_T}{\partial v_0}. \tag{4.4.18}$$

Variation in Ballistics Coefficient

Employing relations (4.4.7) and (4.4.8), we find correction Δc in the ballistics coefficient c related with a small change in the departure angle $\Delta \alpha_0$:

$$\Delta c = \frac{\cos(2\alpha_0)}{\cos^2 \alpha_0 \tan|\alpha_T|(\tan \alpha_0 \,/\, \tan|\alpha_T|-1)} \cdot c \cdot \Delta \alpha_0. \tag{4.4.19}$$

Total Change in Range

Using the above formulae, we can find the change in the horizontal range that corresponds to the change of any of the factors presented above.

For example, the correction ΔX_T in the horizontal range that corresponds only to a small change ΔT_c in the temperature of the propellant charge is

---

[80]   According to a private communication with Dr. Jan Krčmář (Czech Republic)

$$\Delta x_T = 0.001 \cdot v_0 \frac{\partial x_T}{\partial v_0} \Delta T_c .$$

The correction in range that corresponds to a small change Δv_0 in initial speed is

$$\Delta x_T = [\frac{v_0}{\beta \cdot c \cdot G_D(v_0)}(1 + \frac{\tan \alpha_0}{\tan|\alpha_T|} - \frac{g x_T}{v_0^2 \cos^2 \alpha_0 \tan \alpha_T})] \cdot \Delta v_0 .$$

The total horizontal correction related with each small change,

$$\Delta v_0 , \ \Delta \alpha_0 , \ \Delta c , \ \Delta m , \ \Delta \tau_0 , \ \Delta p_0 , \ \Delta T_c , \ \text{etc.},$$

can be founded substituting in (4.4.2) each correction coefficient, determined by (4.4.6)-(4.4.18).

Ballistics Wind Corrections

The range-wind component w_R of the ballistics wind w, deviates the terminal point along the direction of flight (x-axis), while the cross-wind component w_C deviates the terminal point to the right or to the left of the firing plane (in the direction of z-axis).

Range-wind, changes the departure speed of projectile and the departure angle.

- The change in the horizontal range, caused by the range-wind speed w_R, is

$$\Delta x_{wR} = w_R [t - \frac{\partial(x_T)}{\partial v_0} \cdot \cos \alpha_0 + \frac{\partial(x_T) \cdot \sin \alpha_0}{\partial \alpha_0} \frac{1}{v_0}], \qquad (4.4.20)$$

where

$$\frac{\partial x_T}{\partial v_0} = \frac{v_0}{\beta \cdot c \cdot G_D(v_0)}(1 + \frac{\tan \alpha_0}{\tan|\alpha_T|} - \frac{g x_T}{v_0^2 \cos^2 \alpha_0 \tan|\alpha_T|}) \qquad (4.4.21)$$

and

$$\frac{\partial x_T}{\partial \alpha_0} = \frac{x_T \cos(2\alpha_0)}{\cos^2 \alpha_0 \tan|\alpha_T|} \qquad (4.4.22)$$

are the correction coefficients that correspond respectively to the range-wind and cross-wind.

- The deviation of the impact point in the perpendicular direction to the departure plain of the projectile, i.e., in the positive direction of the z-axis, or, in opposite, is

$$z_{Wc} = w_c (t_T - \frac{x_T}{v_0 \cos \alpha_0}). \qquad (4.4.23)$$

The other corrections related with the rotation of the projectile, rotation of Earth, etc., will be briefly seen at the end of the chapter.

Correction in Departure Angle

The relation

$$\Delta \alpha_0 = - \frac{\partial x_T / \partial v_0}{\partial x_T / \partial \alpha_0} \cdot \Delta v_0, \qquad (4.4.24)$$

$$(\frac{\partial x_T}{\partial v_0} = \frac{v_0}{\beta \cdot c \cdot G_D(v_0)} (1 + \frac{\tan \alpha_0}{\tan|\alpha_T|} - \frac{gx_T}{v_0^2 \cos^2 \alpha_0 \tan|\alpha_T|}),$$

$$\frac{\partial x_T}{\partial \alpha_0} = \frac{x_T \cos(2\alpha_0)}{\cos^2 \alpha_0 \tan|\alpha_T|}),$$

can be used to find the correction in angle $\Delta \alpha_0$ that is needed to hit the target and correct the trajectory of projectile flight when there is a change Δv_0 in the departure speed. That formula can be used to study the influence of the variation in the departure angle caused by a small change in the initial speed of the projectile.

The relation (4.4.19) can be easily obtained considering the following relations

$$\Delta x_{T\alpha} = \frac{\partial x_T}{\partial \alpha_0} \Delta \alpha_0 \text{ and } \Delta x_{Tv} = \frac{\partial x_T}{\partial v_0} \Delta v_0 \qquad (4.4.25)$$

that express the change in range respectively for small changes in departure speed and departure angle.

Assuming that the left sides of the equations (4.20) are equal, we can write

$$\frac{\partial x_T}{\partial \alpha_0} \Delta \alpha_0 = \frac{\partial x_T}{\partial v_0} \Delta v_0.$$

Solving the above equation for $\Delta \alpha_0$, we obtain (4.4.19). The minus sign in (4.4.19) is used to compensate for the change in range.

Adjustment of Shooting

Standard range tables are constructed to be used in practice to set up the departure angle in order to strike on the target located at a given horizontal range in standard atmosphere considering that the ballistics characteristics of the projectile are standard.

Because of the variations in atmospheric characteristics and in ballistics characteristics of the projectile, the required standard characteristics are rarely met in practice of shooting. The variation of at least one of the required standard characteristics causes the projectile to miss the target located at a certain range from the muzzle; the point of impact is Δx_T meters in front of the target or on the back.

The range corrections included in the standard range tables allow us to adjust the departure angle by a quantity $\Delta \alpha_0$ in order to compensate for the deviation Δx_T of the striking point from the target.

The correction in departure angle $\Delta \alpha_0$ that is needed to compensate for the total change in range Δx_T is

$$\Delta \alpha_{0T} = - \frac{\Delta x_T}{(\partial x_T / \partial \alpha_0)} , \quad \text{or} \quad \Delta \alpha_{0T} = - \frac{\partial \alpha_0}{\partial x_T} \Delta x_T , \qquad (4.4.26)$$

where

$$\Delta x_T = \frac{\partial x_T}{\partial v_0} \Delta v_0 + \frac{\partial x_T}{\partial \alpha_0} \Delta \alpha_0 + \frac{\partial x_T}{\partial c} \Delta c + \frac{\partial x_T}{\partial \tau_0} \Delta \tau_0 + ... \quad (4.4.27)$$

The examples following this section illustrate the Siacci's method to compute the corrections in range and in departure angle.

The correction theory and particularly Siacci's method have many theoretical and practical applications, especially in exterior ballistics of small arms.

Example 4.1 Main Correction Coefficients

Estimate the main correction coefficients that correspond to the standard horizontal range $x_T = 11000m$ reached when a 122mm Russian projectile (projectile mass of $m = 27.3kg$, mass of the propellant charge $m_P = 3kg$) is fired with the standard initial speed $v_0 = 885m/s$ at an angle of $\alpha_0 = 7.31667°$ in a TSA-standard atmosphere.

The other elements of the trajectory at the given horizontal range 11000m are: time of flight, $t = 20s$; terminal angle, $\alpha_T = -13°$; maximum altitude, $x_m = 463m$; terminal speed, $v_T = 383m/s$.

The ballistics coefficient related with the G_{43}-function of resistance (G-function of the year 1943), as a function of departure angle, $c = f(\alpha_0)$, is given in the following table.[81]

Table 1—The BC of 122mm projectile; departure speed, 885m/s.

Angle, α_0	$\leq 5°$	15°	25°	35°	45°
BC, $c = c(\alpha_0)$	0.496	0.514	0.516	0.519	0.521

81.   "*Range Tables of 122mm cannon Model 1960* ", p. 117, Ministry of Defense of Albania, 1967

The intermediate values can be found by interpolation.

(a) Find the main ballistics coefficients that correspond respectively only to: the initial speed, only to the departure angle, or only in the ballistics coefficient.

(b) Find the change in horizontal range that corresponds only to:

- an increase of 1% in the initial speed i.e. an increase of $\Delta v_0 = 1\% v_0 = 8.85 m / s$.
- an increase of $\Delta \alpha_0 = 0.06°$ in the departure angle.
- an increase of 1% in the ballistics coefficient, i.e., an increase of $\Delta c = 1\% \cdot c = 0.005 m^2 / kg$.

(c) Find the correction in departure angle and the "corrected departure angle" (needed to hit the target) if there is a change of -1.4% in the departure speed, i.e., $\Delta v_0 = - 1.4\% v_0 = - 12.39 m / s$.

(d) Find the correction in departure angle and the "corrected departure angle" that is needed to hit the target located at the range of 11,000 meters considering at the same time a change in departure speed of $\Delta v_0 = 8.85 m / s$ and a change in departure angle of $\Delta \alpha_0 = 0.06°$.

(e) Find the correction coefficients, related with the changes in the ballistics characteristics of the projectile and the propellant charge when there is an increase/decrease of 1% in: projectile mass, propellant temperature, and propellant mass):

(f) Find the correction in departure angle that corresponds to an increase of (2/3)% in the projectile mass, an increase of 10 degrees in the temperature of propellant, and an increase (1/3)% in the propellant mass;, i.e., respectively 0.18kg, 10 degrees, and 0.01kg.

(g) Find the corrections in the horizontal range and in the direction of the z-axis (perpendicular to the departure plane) if there is a range-wind of +10m/s and a cross wind of 10m/s in the direction of z-axis.

Solution:

Preparatory data

The G_{43}-function, for speeds greater than 256m/s, is

$$G_D(v) = 0.157713 \cdot v - 36.39542. \tag{1}$$

Substituting in (1) the initial speed $v_0 = 885m/s$, we find

$$G_D(v_0) = 0.157713 \cdot (885) - 36.39542 = 103.1806$$

Using table 1, by interpolation, we find that the ballistics coefficient of the given projectile is
$$c = f(7.31667°) = 0.50017. \tag{2}$$

We find as well:

$$\bar{y} = (2/3)y_m = (2/3) \cdot (463) = 308.7, \tag{3}$$

$$h(\bar{y}) = (\frac{\tau_{ON} - 0.006328\bar{y}}{\tau_{ON}})^{4.4} = [\frac{289.08 - 0.006328 \cdot (308.7)}{289.08}]^{4.4} = 0.9706, \tag{4}$$

$$\beta = \frac{h(\bar{y})}{\sqrt{\cos\alpha_0}} = \frac{0.9706}{\sqrt{\cos(7.31667)}} = 0.97459. \tag{5}$$

(a) Correction coefficient related with the projectile departure speed

$$\frac{\partial x_T}{\partial v_0} = \frac{v_0}{\beta \cdot c \cdot G_D(v_0)}(1 + \frac{\tan\alpha_0}{\tan|\alpha_T|} - \frac{gx_T}{v_0^2 \cos^2\alpha_0 \tan|\alpha_T|}) =$$

$$= \frac{885}{(0.97459) \cdot (0.50017) \cdot (103.18)}(1 + \frac{\tan(7.31667°)}{\tan|-13°|} -$$

$$- \frac{(9.80665) \cdot (11000)}{(885)^2 \cos^2(7.31667°) \cdot \tan|-13°|}) = 16.71\frac{m}{m/s}$$

Correction coefficient corresponding to the departure angle

$$\frac{\partial x_T}{\partial \alpha_0} = \frac{x_T \cos(2\alpha_0)}{\cos^2 \alpha_0 \cdot \tan|\alpha_T|} = \frac{11000 \cdot \cos(2 \cdot 7.31667)}{\cos^2(7.31667) \cdot \tan|-13|} = 46860.5 \frac{m}{radian} .$$

Correction coefficient corresponding to the ballistics coefficient

$$\frac{\partial x_T}{\partial c} = -(1 - \frac{\tan \alpha_0}{\tan|\alpha_T|}) \frac{x_T}{c} = -(1 - \frac{\tan(7.31667)}{\tan|-13|}) \cdot \frac{11000}{0.50017} = -9750.92 \frac{m}{m^2 / kg} .$$

(b) Horizontal range correction that corresponds to a change of $\Delta v_0 = 8.85m/s$ is

$$\Delta x_{Tv} = \frac{\partial x_T}{\partial v_0} \cdot \Delta v_0 = (16.71) \cdot (8.85) = 147.88m . \quad (7)$$

The projectile will exceed the target located 11,000 meters from the muzzle, and will hit the ground 11,147.88 meters from the muzzle of the cannon.

An increase of $\Delta \alpha_0 = 0.06°$ in the departure angle only increases the horizontal range by

$$\Delta x_{T\alpha} = \frac{\partial x_T}{\partial \alpha_0} \Delta \alpha_0 = (46860.5) \cdot (\frac{\pi}{180}) \cdot (0.06) = 49.07m . \quad (8)$$

An increase of $0.005m^2/kg$ in the ballistics coefficient will decrease the projectile range by 48.77m:

$$\Delta x_{Tc} = \frac{\partial x_T}{\partial c} \Delta c = (-9750.92m) \cdot (0.005) = -48.77m . \quad (9)$$

(c) A decrease of $\Delta v_0 = -12.39m/s$ in the departure speed decreases the projectile range by

$$\Delta x_{Tv} = \frac{\partial x_T}{\partial v_0} \cdot \Delta v_0 = (16.71) \cdot (-12.39) = -207.04m \quad (10)$$

and the departure angle by

$$\Delta \alpha_0 = -\frac{\partial x_T / \partial v_0}{\partial x_T / \partial \alpha_0} \cdot \Delta v_0 = -\frac{16.71}{46860.5(\pi / 180)}(-12.39) = +0.25314° . \quad (11)$$

In order to correct the shooting, i.e., to hit the target located at the horizontal range of 11,000 meters, we need to increase the departure angle by $\Delta \alpha_0 = 0.25314°$.

The corrected departure angle is

$$\alpha_0 + \Delta \alpha_0 = 7.31667° + 0.25314° = 7.56981° .$$

(d) Using (8) and (9), we find that the total change in range that corresponds to changes

$$\Delta v_0 = 8.85 m / s \text{ and } \Delta \alpha_0 = 0.06° \text{ is}$$

$$\Delta x_T = \Delta x_{Tv} + \Delta x_{T\alpha} = 147.88 + 49.07 = 196.95 m .$$

Substituting in (4.4.21), we find the correction in angle is

$$\Delta \alpha_{0T} = -\frac{\Delta x_T}{(\partial x_T / \partial \alpha_0)} = -\frac{196.95}{817.87} = -0.2408° .$$

The adjusted angle is

$$\alpha_0 + \Delta \alpha_{0T} = 7.31667° - 0.2408° = 7.07586° .$$

(e) The correction coefficient related with the increase in the projectile mass is

$$\frac{\partial x_T}{\partial m} = (1 - \frac{\tan \alpha_0}{\tan |\alpha_T|} - 0.4 \cdot \frac{v_0}{x_T} \frac{\partial x_T}{\partial v_0}) \frac{x_T}{m} =$$

$$= [1 - \frac{\tan(7.31667°)}{\tan |-13°|} - 0.4 \cdot \frac{885}{11000} \cdot (16.71)] \cdot \frac{11000}{3} = -344.35.$$

The correction coefficient that corresponds to the temperature (T_c) of the propellant charge is

$$\frac{\partial x_T}{\partial T_c} = 0.00128 \cdot v_0 \frac{\partial x_T}{\partial v_0} = 0.00128 \cdot (885) \cdot (16.71) = 18.93 \, .$$

The correction coefficient that corresponds to the mass (m_P) of the propellant charge is

$$\frac{\partial x_T}{\partial m_P} = 0.60 \cdot v_0 \frac{\partial x_T}{\partial v_0} = 0.60 \cdot (885) \cdot 16.71 = 8873 \, .$$

(f) The change in the horizontal range is

$$\Delta x_T = \frac{\partial x_T}{\partial m} \Delta m + \frac{\partial x_T}{\partial T_c} \Delta T_c + \frac{\partial x_T}{\partial m_P} \Delta m_P =$$

$$= (-344.35) \cdot (0.18) + (18.93) \cdot (10) + (8873) \cdot (0.01) = 216m.$$

The change in the departure angle is

$$\Delta \alpha_{0T} = - \frac{\Delta x_T}{(\partial x_T / \partial \alpha_0)} = - \frac{216}{817.87} = -0.2642° \, .$$

The adjusted departure angle is

$$\alpha_0 + \Delta \alpha_{0T} = 7.31667° - 0.2642° = 7.0525° \, .$$

(g) Using (4.4.15), we find that the increase in the horizontal range that corresponds to the range wind +10m/s is

$$\Delta x_{wR} = w_R [t - \frac{\partial(x_T)}{\partial v_0} \cdot \cos \alpha_0 + \frac{\partial(x_T) \cdot \sin \alpha_0}{\partial \alpha_0} \cdot \frac{1}{v_0}] =$$

$$= 10 \cdot [20 - 16.71 \cdot \cos(7.31667°) + 46860.5 \cdot \frac{\sin(7.31667)}{885}] = 101.7m.$$

The deflection of the point of impact in the positive direction of z-axis is

$$z_{Wc} = w_c \left(t_T - \frac{x_T}{v_0 \cos\alpha_0}\right) = 10 \cdot \left(20 - \frac{11000}{885 \cdot \cos(7.31667)}\right) = 74.7m.$$

Example 4.2 Projectile Trajectory in Non-standard Atmosphere

Use the Siacci's method to solve the example 1.3, section 4.1, or example 2.1, section 4.2, that are reformulated hereafter:

The elements of the trajectory of a 122mm projectile for the horizontal range $x_T = 10000m$, in TSA atmosphere (virtual temperature $\tau_{0N} = 289.08° K$ (temperature of dry air, 15 degrees Celsius; relative humidity of air, 50%; pressure of water vapors, 6.35mm Hg.) are:
departure angle is $\alpha_{01} = 6.2°$; time of flight to the target is $t_T = 17s$; terminal angle is $\alpha_T = 10°$; trajectory maximum altitude is $y_m = 345m$.

(a) Find the departure angle needed to hit the target located at the horizontal range of 10000 meters if temperature of air is 25 degrees Celsius, and the pressure is 740 mm Hg. Consider the air humidity as 80%.
(b) Find as well the change in horizontal range that corresponds to a change in propellant temperature, and the correction angle needed to hit the target 10,000 meters away, considering all the effects of the temperature and pressure.

Solution:

Preparatory Data

In example 1.3, we have found that the virtual temperature is $\tau = 301.08° K$. Thus, the change in the virtual temperature and the change in pressure are respectively

$$\Delta\tau_0 = 301.08° - 289.08° = 12.72° K \quad \text{and}$$
$$\Delta p_0 = 740 - 750 = -10mmHg$$

As in example 4.1, we find

$$G_D(v_0) = 0.157713 \cdot (885) - 36.39542 = 103.1806$$

(a) Using table 1, example 4.1, by interpolation, we find that the ballistics coefficient of the given projectile is

$$c = f(6.2°) = 0.49816. \tag{2}$$

We find as well:

$$\bar{y} = (2/3)y_m = (2/3) \cdot (345) = 230, \tag{3}$$

$$h(\bar{y}) = (\frac{\tau_{0N} - 0.006328\bar{y}}{\tau_{0N}})^{4.4} = [\frac{289.08 - 0.006328 \cdot (230)}{289.08}]^{4.4} = 0.978036, \tag{4}$$

$$\beta = \frac{h(\bar{y})}{\sqrt{\cos \alpha_0}} = \frac{0.9706}{\sqrt{\cos(6.2)}} = 0.98081. \tag{5}$$

The correction coefficient related with the projectile departure speed is

$$\frac{\partial x_T}{\partial v_0} = \frac{v_0}{\beta \cdot c \cdot G_D(v_0)}(1 + \frac{\tan \alpha_0}{\tan|\alpha_T|} - \frac{gx_T}{v_0^2 \cos^2 \alpha_0 \tan|\alpha_T|}) =$$

$$= \frac{885}{(0.98091) \cdot (0.49816) \cdot (103.18)}(1 + \frac{\tan(6.2°)}{\tan|-10°|} -$$

$$- \frac{(9.80665) \cdot (10000)}{(885)^2 \cos^2(6.2°) \cdot \tan|-10°|}) = 15.76.$$

The correction coefficient related with the projectile departure angle is

$$\frac{\partial x_T}{\partial \alpha_0} = \frac{x_T \cos(2\alpha_0)}{\cos^2 \alpha_0 \cdot \tan|\alpha_T|} = \frac{10000 \cdot \cos(2 \cdot 6.2)}{\cos^2(6.2) \cdot \tan|-10|} = 57046.54 \frac{m}{radian}.$$

The correction coefficient expressed in meter/degree is

$$\frac{\partial x_T}{\partial \alpha_0} = 57046 .54 \cdot (\frac{\pi}{180}) = 995 .65 \frac{m}{deg\, ree} .$$

The correction coefficient that corresponds to the temperature of air is

$$\frac{\partial x_T}{\partial \tau_0} = (1 - \frac{v_0}{2x_T} \frac{\partial x_T}{\partial v_0}) \frac{x_T}{\tau_0} = (1 - \frac{885}{2 \cdot 10000} \cdot 15.76) \cdot \frac{10000}{289.08} = 10.47 .$$

The correction coefficient that corresponds to the atmospheric pressure is

$$\frac{\partial x_T}{\partial p_0} = -(1 - \frac{\tan \alpha_0}{\tan|\alpha_T|}) \frac{x_T}{p_0} = -(1 - \frac{\tan(6.2)}{\tan|-10|}) \cdot \frac{10000}{750} = -5.12 .$$

The total change in range is

$$\Delta x_T = \frac{\partial x_T}{\partial \tau_0} \Delta \tau_0 + \frac{\partial x_T}{\partial p_0} \Delta p_0 = (10.47) \cdot (10) + (-5.12) \cdot (-10) = 155.9 m .$$

The correction in departure angle is

$$\Delta \alpha_{0T} = -\frac{\Delta x_T}{(\partial x_T / \partial \alpha_0)} = -\frac{155.9}{995.65} = -0.15658° .$$

In order to correct shooting, i.e., to hit the target located at the horizontal range of 10,000 meters, we need to decrease the departure angle by $\Delta \alpha_{0T} = 0.15658°$.

The corrected departure angle is

$$\alpha_0 + \Delta \alpha_0 = 6.2° - 0.15658° = 6.04312° .$$

(b) The correction coefficient that corresponds to the temperature (T_c) of the propellant charge is

$$\frac{\partial x_T}{\partial T_c} = 0.001285 \cdot v_0 \frac{\partial x_T}{\partial v_0} = 0.001285 \cdot (885) \cdot 15.76 = 17.9 .$$

The corresponding change in the horizontal range is

$$\Delta x_{Tc} = \frac{\partial x_T}{\partial T_c} \Delta \tau_c = 17.9 \cdot 10 = 179m .$$

The total change in range related with the variation in temperature, pressure, and temperature of propellant charge is

$$\Delta x_T = 155.9 + 179 = 335m .$$

The change in departure angle that results from the change of all factors is

$$\Delta \alpha_{0T} = -\frac{\Delta x_T}{(\partial x_T / \partial \alpha_0)} = -\frac{335}{995.65} = -0.33659° .$$

The corrected departure angle is

$$\alpha_0 + \Delta \alpha_0 = 6.2° - 0.33589° = 5.86341° .$$

$$\Delta x_{wR} = w_R [t - \frac{\partial(x_T)}{\partial v_0} \cdot \cos \alpha_0 + \frac{\partial(x_T)}{\partial \alpha_0} \cdot \frac{\sin \alpha_0}{v_0}] =$$

$$= (10) \cdot [1.592 - (1.838) \cdot \cos(0.5978°) + 66433.05 \cdot \frac{\sin(0.5978°)}{830}] = 6m.$$

Example 4.3 Correction Coefficients for the Lapua Bullet

Consider a 338 Lapua GB528 Scenar 19.44g (300 grain) bullet fired with a departure speed of 830m/s, at the sea level in the ICAO atmosphere (pressure, 760 mm Hg; temperature, 15 degrees Celsius; density of air, 1.225kg/m^3; speed of sound, 340.30m/s; relative humidity, 0%).

The elements of the trajectory at the horizontal range of 1000 meters are:

departure angle, 0.5978 degree; time of flight,
1.592s; terminal speed, 479.65m/s; terminal angle, -
0.8623 degree; vertex, (549, 3.12).

The ballistics coefficient of the given projectile with respect to
the G_1-function is $c = 1.840m^2 / kg$ ($C = 0.773lb / in^2$).
Use the Siacci's method to find:

(a) the main correction coefficients.
(b) Find tthe total change in horizontal range that corresponds
to:
- an increase of 1% in the initial speed, i.e. an increase of
$\Delta v_0 = 1\%v_0 = 8.30m / s$.
- an increase of $\Delta \alpha_0 = 0.06°$ in the departure angle.
- an increase of 1% in the ballistics coefficient, i.e., an
increase of
$\Delta c = 1\% \cdot c = 0.0184m^2 / kg$.

Find as well the corresponding correction in angle and the
corrected angle that corresponds to the total change in the horizontal
range.

(c) Find the total change in horizontal range that corresponds
to:
- an increase of 10 degrees in the in temperature of air
- an increase of 10 degrees in the temperature of the
propellant
- an increase of 3% (i.e., 0.00058kg) in the projectile
mass

Find as well the corresponding correction in angle and the
corrected angle that corresponds to the total change in the
horizontal range.

(d) Find the correction in horizontal range related with a range-
wind of +10m/s and the deflection of the bullet in the
direction of z-axis related with a cross wind of +10m/s.

Find the correction in the departure angle and the corrected departure angle related with the range wind, as well as the deflection angle.

(e) Find the total change in the horizontal range that results from all changes in (a)-(d). Find as well the total angle correction and the related corrected angle.

(f) Find the correction in range and the correction in departure angle that result from a change of 0.05 meters in the barrel length.

Solution:

The G_1-function of resistance (formula 1.5.4) in ICAO atmosphere, for projectiles flying with speed greater than 256m/s, is

$$G_1(v) = 0.315754 \cdot v - 78.6769. \qquad (1)$$

Substituting in (1) the initial speed $v_0 = 830m/s$, we find

$$G_1(v_0) = 0.315754 \cdot (830) - 78.6769 = 183.40.$$

Since the trajectory is very low, we can consider $h(\bar{y}) = 1$. Thus,

$$\beta = \frac{h(\bar{y})}{\sqrt{\cos\alpha_0}} = \frac{1}{\sqrt{\cos(0.5978)}} = 1.$$

(a) Correction coefficient related with the projectile departure speed

$$\frac{\partial x_T}{\partial v_0} = \frac{v_0}{\beta \cdot c \cdot G_D(v_0)}(1 + \frac{\tan\alpha_0}{\tan|\alpha_T|} - \frac{gx_T}{v_0^2 \cos^2(\alpha_0)\tan|\alpha_T|}) =$$

$$= \frac{830}{(1)\cdot(1.84)\cdot(183.4)}(1 + \frac{\tan(0.5978°)}{\tan|-0.8623°|} -$$

$$- \frac{(9.80665)\cdot(1000)}{(830)^2 \cos^2(0.5978°)\cdot\tan|-0.8623°|}) = 1.838\frac{m}{m/s}.$$

The correction coefficient related with the projectile departure angle is

$$\frac{\partial x_T}{\partial \alpha_0} = \frac{x_T \cos(2\alpha_0)}{\cos^2 \alpha_0 \cdot \tan|\alpha_T|} = \frac{1000 \cdot \cos(2 \cdot 0.5978)}{\cos^2(0.5978) \cdot \tan|-0.8623|}) = 66433.05 \frac{m}{radian} \cdot$$

The correction coefficient related with the projectile ballistics coefficient is

$$\frac{\partial x_T}{\partial c} = -(1 - \frac{\tan \alpha_0}{\tan|\alpha_T|}) \frac{x_T}{c} = -(1 - \frac{\tan(0.5978)}{\tan|-0.8623|}) \cdot \frac{1000}{1.840} = -166.72 \frac{m}{m^2 / kg} \cdot$$

(b) The total change in the horizontal range is

$$\Delta x_T = \frac{\partial x_T}{\partial v_0} \Delta v_0 + \frac{\partial x_T}{\partial \alpha_0} \Delta \alpha_0 + \frac{\partial x_T}{\partial c} \Delta c =$$

$$= (1.838) \cdot (8.30) + (66433.05) \cdot (\frac{\pi}{180}) \cdot (0.06) + (-166.72) \cdot (0.0184) =$$

$$= 15.26 + 69.57 - 3.07 = 81.75m.$$

Note that the change in the horizontal range related with the given change in ballistics coefficient is very small (-3.07m), and can be neglected.

The departure angle correction is

$$\Delta \alpha_{0T} = -\frac{\Delta x_T}{(\partial x_T / \partial \alpha_0)} = -\frac{81.75}{66433.05(\pi / 180)} = -0.0705° \cdot$$

The corrected departure angle is

$$\alpha_0 + \Delta \alpha_{0T} = 0.5978° - 0.0705° = 0.5273° \cdot$$

(c) Since the relative humidity is 0%, the virtual temperature is equal to the temperature of air in degree Kelvin that corresponds to 25 degrees Celsius, i.e.,

$$T_0 = 273.15 + 25 = 298.15° \ K$$

The correction coefficient related with the change in temperature of air is

$$\frac{\partial x_T}{\partial \tau_0} = (1 - \frac{v_0}{2x_T} \frac{\partial x_T}{\partial v_0}) \frac{x_T}{\tau_0} = (1 - \frac{830}{2 \cdot (1000)} \cdot 1.838) \cdot \frac{1000}{298.15} = 0.796 \frac{m}{° \ K} \ .$$

The correction coefficient related with the change in temperature of propellant charge (see 4.4.13b) is

$$\frac{\partial x_T}{\partial T_c} = 0.0014 \cdot v_0 \frac{\partial x_T}{\partial v_0} = 0.0014 \cdot (830) \cdot (1.838) = 2.136 \frac{m}{° \ K} \ .$$

The correction coefficient related with the change in the projectile mass is

$$\frac{\partial x_T}{\partial m} = (1 - \frac{\tan \alpha_0}{\tan|\alpha_T|} - 0.4 \cdot \frac{v_0}{x_T} \frac{\partial x}{\partial v_0}) \frac{x_T}{m} =$$

$$= [1 - \frac{\tan(0.5978°)}{\tan|-0.8623|} - 0.4 \cdot \frac{830}{1000} \cdot (1.838)] \cdot \frac{1000}{0.01944} = -15609.62 \frac{m}{kg} \ .$$

The total correction in the horizontal range is

$$\Delta x_T = \frac{\partial x_T}{\partial \tau_0} \Delta \tau_0 + \frac{\partial x_T}{\partial T_C} \Delta T_C + \frac{\partial x_T}{\partial m} \Delta m =$$

$$= (0.796) \cdot (10) + (2.136) \cdot (10) + (-15609.62) \cdot (0.00058) =$$

$$= 7.96 + 21.36 - 9.05 = 20.27m.$$

The correction in angle is

$$\Delta \alpha_{0T} = - \frac{\Delta x_T}{(\partial x_T / \partial \alpha_0)} = - \frac{20.27}{66483.05 \cdot (\pi / 180)} = -0.0175° \ .$$

The corrected departure angle is

$$\alpha_0 + \Delta \alpha_{0T} = 0.5978° - 0.0175° = 0.5803° \ .$$

(d) The change in the range that corresponds to a range-wind of +10m/s is

$$\Delta x_{wR} = w_R [t - \frac{\partial(x_T)}{\partial v_0} \cdot \cos\alpha_0 + \frac{\partial(x_T)}{\partial\alpha_0} \cdot \frac{\sin\alpha_0}{v_0}] =$$

$$= (10) \cdot [1.592 - (1.838) \cdot \cos(0.5978°) + 66433.05 \cdot \frac{\sin(0.5978°)}{830}] = 6m.$$

The correction in departure angle is

$$\Delta\alpha_{0T} = -\frac{\Delta x_T}{(\partial x_T / \partial\alpha_0)} = -\frac{6}{66483.05 \cdot (\pi/180)} = -0.005°.$$

The cross-wind deflection in the positive direction of z-axis is

$$z_{Wc} = w_c(t_T - \frac{x_T}{v_0\cos\alpha_0}) = 10 \cdot (1.592 - \frac{1000}{830 \cdot \cos(0.5978)}) = 3.871m$$

$$\alpha_D = ar\tan(3.871/1000) = 0.2218°.$$

The correction in departure angle is

$$\Delta\alpha_{0T} = -\frac{\Delta x_T}{(\partial x_T / \partial\alpha_0)} = -\frac{6}{66483.05 \cdot (\pi/180)} = -0.005°.$$

The corrected angle is

$$\alpha_0 + \Delta\alpha_{0T} = 0.5978° - 0.005° = 0.5928°.$$

The cross-wind deflection in the positive direction of z-axis is

$$z_{Wc} = w_c(t_T - \frac{x_T}{v_0\cos\alpha_0}) = 10 \cdot (1.592 - \frac{1000}{830 \cdot \cos(0.5978)}) = 3.871m.$$

The deflection angle is

$$\alpha_D = ar\tan(3.871/1000) = 0.2218°.$$

(e) The total change in range that results from all changes in (b), (c), and (d) is

$$\Delta x_T = (81.75) + (20.27) + (6) = 108.2m.$$

Correction in angle is

$$\Delta \alpha_{0T} = -\frac{\Delta x_T}{(\partial x_T / \partial \alpha_0)} = -\frac{108.2}{66483.05 \cdot (\pi / 180)} = -0.0934°$$

The corrected departure angle is

$$\alpha_0 + \Delta \alpha_{0T} = 0.5978° - 0.0934° = 0.5044°.$$

(f) The correction coefficient related with the barrel length is

$$\frac{\partial x_T}{\partial L_B} = 0.28916 \cdot v_0 \frac{\partial x_T}{\partial v_0} = 0.28916 \cdot (830) \cdot (1.838) = 441.125m$$

The change in the horizontal range is

$$\Delta x_T = \frac{\partial x_T}{\partial L} \Delta L = (441.125) \cdot (-0.05) = -22.056m.$$

The correction in angle is

$$\Delta \alpha_{0T} = -\frac{\Delta x_T}{(\partial x_T / \partial \alpha_0)} = -\frac{-22.056}{66483.05 \cdot (\pi / 180)} = -0.01904°.$$

The corrected angle is

$$\alpha_0 + \Delta \alpha_{0T} = 0.5978° + 0.01904° = 0.61684°.$$

4.5. General Approach to Range Corrections

Field Artillery

For field artillery shooting, the main correction coefficients

$$\frac{\partial x_T}{\partial v_0}, \frac{\partial x_T}{\partial \alpha_0}, \frac{\partial x_T}{\partial c}, \qquad (4.5.1)$$

can be found using the approximate derivatives:

$$\frac{\partial x_T}{\partial v_0} \approx \frac{x_T(v_0 + \Delta v_0) - x_T(v_0 - \Delta v_0)}{2 \cdot \Delta v_0}, \qquad (4.5.2)$$

$$\frac{\partial x_T}{\partial v_0} \approx \frac{x_T(v_0 + \Delta v_0) - x_T(v_0 - \Delta v_0)}{2 \cdot \Delta v_0} \qquad (4.5.3)$$

and

$$\frac{\partial x_T}{\partial c} \approx \frac{x_T(c + \Delta c) - x_T(c - \Delta c)}{2 \cdot \Delta c} \qquad (4.5.4)$$

The horizontal range x_T in (4.5.2)-(4.5.4) is displayed in the range tables of the firearm that is constructed, for example, using the method presented in section 2.21:

The horizontal range x_T and all the corresponding elements of the trajectory can be determined using the numerical integration of the differential equations of the projectile flight, the ballistics coefficient function given by the manufacturer, and using the PC programs BCPROJ.BAS, APROJC.BAS, and RPROJC.BAS.

Once the horizontal range x_T is calculated, we are able to find the main correction coefficients using (4.5.2)-(4.5.4).

After the main correction coefficients are determined for each horizontal range, we can use the formulas (4.4.10)-(4.4.18) to determine the secondary correction coefficients.

The example 5.1 illustrates the use of the PC programs to calculate the main correction coefficients based on the formulas presented in this section.

Antiaircraft Artillery

The above formulas can be used to find the main correction coefficients for the inclined fire, or antiaircraft artillery.

For the target located on the inclined plane, we have to find the main correction coefficient for x-coordinate as well the main correction coefficient for the y-coordinate of the target, i.e.,

$$(\frac{\partial x_T}{\partial v_0}, \frac{\partial y_T}{\partial v_0}), \quad (\frac{\partial x_T}{\partial \alpha_0}, \frac{\partial y_T}{\partial \alpha_0}), \quad (\frac{\partial x_T}{\partial c}, \frac{\partial y_T}{\partial c}). \qquad (4.5.5)$$

For the x-coordinate, the main correction coefficient already are determined using (4.5.2)-(4.5.4). Similar main correction coefficient formulae can be written for the y-coordinate.

The correction in horizontal direction and the correction in the vertical direction of firing that result from the influence of range wind can be estimated using the following formulae [82]:

$$\Delta x_T = w_R (t - \frac{\partial x_T}{\partial v_0} \cos \alpha_0 + \frac{\partial x_T}{\partial \alpha_0} \frac{\sin \alpha_0}{v_0}), \qquad (4.5.6)$$

$$\Delta y_T = w_R (t - \frac{\partial y_T}{\partial v_0} \cos \alpha_0 + \frac{\partial y_T}{\partial \alpha_0} \frac{\sin \alpha_0}{v_0}). \qquad (4.5.7)$$

Example 5.1 Main Correction Coefficients

Reference: Example 4.1.
(We suppose that we have already constructed the standard range table in TSA atmosphere, for example, using the approach presented in section 2.21).

Estimate the main correction coefficients that correspond to the standard horizontal range $x_T = 11000m$ reached when a 122mm Russian projectile (projectile mass is $m = 27.3kg$,; mass of the propellant charge is $m_p = 3kg$) is fired with the standard initial speed of $v_0 = 885m/s$ at an angle of $\alpha_0 = 7.31667°$ in TSA atmosphere.

The other elements of the trajectory at the given terminal point are: time of flight, $t = 20s$; terminal angle, $\alpha_T = 13°$; maximum altitude, $x_m = 463m$; terminal speed, $v_T = 383m/s$.

---

82. Shapiro, J. M.., *Exterior Ballistics*, p.208, Oborongiz, Moscow, 50'

The experimental ballistics coefficient, $c = f(\alpha_0)$, related with the G_{43}-function of resistance (G-function of the year 1943), is given in the following table.[83]

Table 1: The BC of 122mm projectile; departure speed, 885m/s.

Angle, α_0	≤5°	15°	25°	35°	45°
BC, $c = f(\alpha_0)$	0.496	0.514	0.516	0.519	0.521

The intermediate values can be found by interpolation.

Find the main ballistics coefficients that correspond respectively:

- only to the initial speed,
- only to the departure angle ,
- only to the ballistics coefficient.

Find the change in the horizontal range that corresponds only to:

- An increase of 1% in the initial speed, i.e., an increase of $\Delta v_0 = 1\% v_0 = 8.85 m/s$.
- An increase in the departure angle of $\Delta \alpha_0 = 0.06°$.
- An increase of 1% in the ballistics coefficient, i.e., an increase of $\Delta c = 1\% \cdot c = 0.005 m^2 / kg$.

Find the correction in the departure angle and the "corrected departure angle" (needed to hit the target) if there is a change of - 1.4% in the departure speed, i.e., $\Delta v_0 = -1.4\% v_0 = -12.39 m/s$.

Find the correction in departure angle and the "corrected departure angle" that is needed to hit the target located at the range of 11,000 meters considering at the same time a change in departure speed of $\Delta v_0 = 8.85 m/s$ and a change in departure angle of $\Delta \alpha_0 = 0.06°$.

---

83. "Range Tables of 122mm cannon Model 1960 ", p. 117, Ministry of Defense of Albania, 1967

Solution

Preparatory Data

From table 1, by interpolation, we find that the ballistics coefficient that corresponds to the departure angle $\alpha_0 = 7.31667°$ and $x_T = 11000m$ is

$$c(7.31667) = 0.5.$$

Using the PC program RPROJC.BAS, we find that for the departure angle $\alpha_0 = 7.31667°$, the horizontal range is 10,972, and not $x_T = 11000m$.

It means that the approximate coefficient is not accurate.
To have more accurate data, we consider

$$c(7.31667) = 0.497$$

obtained using PC program BCPROJ.BAS.

Correction Coefficient Related with Departure Speed

Employing the PC program RPROJC.BAS, we find the following data:

$$x_T(v_0 + \Delta v_0) = x(885 + 5) = 11080m \quad \text{and}$$
$$x_T(v_0 - \Delta v_0) = x(885 - 5) = 10911m.$$

The correction coefficient related with the departure speed is

$$\frac{\partial x_T}{\partial v_0} \approx \frac{x_T(v_0 + \Delta v_0) - x_T(v_0 - \Delta v_0)}{2 \cdot \Delta v_0} = \frac{11080 - 10911}{2 \cdot 5} = 16.9.$$

Correction Coefficient Related with the Departure Angle

Using the PC program RPROJC.BAS, we find:

$$x_T(\alpha_0 + \Delta \alpha_0) = x(7.31667° + 0.05°) = 11048m,$$

and

$$x_T(\alpha_0 - \Delta\alpha_0) = x(7.31667° - 0.05°) = 10953m.$$

The correction coefficient related with the departure angle is

$$\frac{\partial x_T}{\partial \alpha_0} \approx \frac{x_T(\alpha_0 + \Delta\alpha_0) - x_T(\alpha_0 - \Delta\alpha_0)}{2 \cdot \Delta\alpha_0} = \frac{11048 - 10953}{2 \cdot 0.05} = 950$$

Correction Coefficient Related with the Ballistics Coefficient

Using the PC program RPROJC.BAS, we find:

$$x_T(c + \Delta c) = x(0.497 + 0.005) = 10948,$$

$$x_T(c - \Delta c) = x(0.497 - 0.005) = 11044$$

and

$$\frac{\partial x_T}{\partial c} \approx \frac{x_T(c + \Delta c) - x_T(c - \Delta c)}{2 \cdot \Delta c} = \frac{10948 - 11044}{2 \cdot 0.005} = 9600.$$

4.6. Estimation of Range Corrections Using Snell's Law

The standard range tables that are constructed for shooting at the sea-level altitude, or close to sea level, are not valid for firing at relatively high altitudes, or for mountain shootings.

For example, the field artillery uses the standard mountain tables constructed for altitudes 500m, 1,000m, 1,500m, 2,000m, 2,500m, and 3,000m.

The standard mountain range tables [84] are constructed in the same way as the sea-level standard tables, for example, using the method shown in section 2.21.

---

84.   Standard mountain range table is called the range table that is constructed based on the characteristics of the atmosphere at the altitude of shooting, when those characteristics are the same as the characteristics that can be obtained from the respective sea-level standard atmosphere using the known laws of the temperature and density, for example the laws shown in formulas (2.2.5), (2.2.6) and (2.2.7).

The correction methods presented in sections 4.4 and 4.6 can be used to find the range corrections and departure angle corrections based on the standard range tables of a firearm.

The correction coefficients obtained based on the sea -level standard range tables can not be applied to find the range corrections for shooting in mountains or high altitudes. For mountain shootings (high-altitude shootings), the correction coefficients, related with small variations of the characteristics of atmosphere and the ballistics characteristics of the projectile/firearm, have to be determined based on the respective standard mountain tables.

Nevertheless, we can use the projectile-streamline model and the Snell's law to determine the range corrections based on the sea - level standard range tables.

In analogy with the correction method introduced in sections 4.1 and 4.2, we present a new approach based on the projectile-streamline and Snell's law model.

We assume that the range corrections related with the standard sea-level range tables are known.

For example, for the sea -level range x_T, we suppose that there are known the range corrections are known:

$$\Delta x_T(v_0), \ \Delta x_T(\alpha_0), \ \Delta x_T(c), \ \Delta x_T(\tau), \ \Delta x_T(p), \ \Delta x_T(T_C), \ \Delta x_T(\alpha_0),$$
$$\Delta x_T(m), \ \Delta x_T(w_r), \ \Delta x_T(w_C),$$

that correspond respectively in the following small changes:

departure speed, $\Delta v_0 = \pm 0.01 v_0$; departure angle, $\Delta \alpha_0 = \pm 0.06°$ [85]; ballistics coefficient, $\Delta c = \pm 0.005 \cdot c$; virtual temperature, $\Delta \tau = \pm 10° C$; atmospheric pressure, $\Delta p = \pm 10 mm \ Hg$; temperature of the propellant charge, $\Delta T_C = 10° C$; projectile mass, $\Delta m = \pm (2/3)\% m$; range-wind, $\Delta w_R = \pm 10 m/s$; cross-wind, $\Delta w_C = \pm 10 m/s$.

---

85. In military to measure angles it is used 1 mil; 1 mil =360/6400 degrees = 0.05625 degrees.

Based on the projectile-similitude relation (4.1.5),

$$X_T = J \cdot x_T, \tag{4.6.1}$$

that exists between ranges of 2two similitude trajectories, for the correction in range ΔX_T that corresponds to X_T, we can write:

$$\Delta X_T = J \cdot \Delta x_T. \tag{4.6.2}$$

For example, the change in the horizontal range that corresponds to a small change in the temperature of air, we have:

$$\Delta X_T(\tau) = J \cdot \Delta x_T(\tau) \tag{4.6.4}$$

The total change in range is

$$\Delta X_T = \Delta X_T(v_0) + \Delta X_T(\alpha_{02}) + ... + \Delta X_T(w_R). \tag{4.6.5}$$

The correction in departure angle $\Delta \alpha_{02}$ that corresponds to the total change in range (4.6.5) is

$$\Delta \alpha_{02} = \frac{\Delta X_T}{\Delta X_T(\alpha_{02})}. \tag{4.4.6}$$

Using the Snell's law, for the correction $\Delta \alpha_{02}$ in the departure angle α_{02} that corresponds to the correction (4.6.2), we can write:

$$\sin \Delta \alpha_{02} = J \cdot \sin \Delta \alpha_0, \tag{4.4.7)}$$

or equivalently,

$$\Delta \alpha_{02} = \arcsin(J \cdot \sin \Delta \alpha_0). \tag{4.4.8}$$

The projectile-similitude and Snell's law approach can be used for the elements of any point on the trajectory of flight. It can be used to find the corrections related with the inclined fire.

Example 6.1 Projectile Trajectory-Streamline and Snell's Law

Correction Method

The departure speed of the 122mm projectile of the Model 1960 cannon model 1960 is reduced by 1%. Use the similitude and Snell's laws and the range table of the respective cannon to find the departure angle needed to hit a target located at the horizontal range of 12,000 meters from the cannon.

In the same way, find the corrections in range that corresponds to:

- a change in propellant temperature of 10 degrees Celsius.
- a change in the range-wind of 10m/s.
- a change in the cross-wind of 10m/s.
- a change in the departure angle of 0.06 degree.

Note that to find the angle and the range in standard conditions, we can use the PC program ACA122.Bas.

Solution:

The similitude horizontal range in standard atmosphere that corresponds to 12,000 meters in the given non-standard atmosphere is

$$x_{T1} = 0.849125 \cdot x_{T2} = 0.849125 \cdot (12000) = 10189.5m \,.$$

In the table below, are given the range corrections for the range 10,189.5 mm are given, as they are shown in the range table of 122mm Russian Mod. 1960 cannon Mod. 1960:

Table 1: Corrections in Range at the Sea Level; similitude range, 10189.5

Correction in range for the change in initial speed of 1%	Correction in range for a change in propellant temperature of 10 degree	Correction in range for a change in range-wind of 10m/s	Correction in range for a change in cross-wind of 10m/s	Correction in range for a change in a departure angle of 0.06 degree
140m	182	84	50m	56

Using the data of table 1, we obtain:

Correction in range at the altitude 1,500 meters that corresponds to a change in initial speed of 1% is

$$\Delta x_{T2} = J \cdot \Delta x_{T1} = 1.17768(140) = 165m$$

In the same way, we find:

The correction for a change in the propellant temperature of 10 degree is

$$\Delta x_{T2} = J \cdot \Delta x_{T1} = 1.17768(182) = 214.3m$$

The correction in range for a change in range-wind of 10m/s is

$$\Delta x_{T2} = J \cdot \Delta x_{T1} = 1.17768(84) = 99m \,.$$

The correction in range for a change in cross-wind of 10m/s is

$$\Delta x_{T2} = J \cdot \Delta x_{T1} = 1.17768(50) = 59m$$

For a change in the departure angle of

$$\Delta \alpha_2 = ar\sin(J \cdot \sin\Delta\alpha_1) = ar\sin[1.17776 \cdot \sin(0.06°)] = 0.0707° \,,$$

the correction in range is

$$\Delta x_{T2} = J \cdot \Delta x_{T1} = 1.17768(56) = 66m \,.$$

From the above results, we find that the change in range that corresponds to a departure angle of 0.06 degrees is 56m.

In the first row of table 2, are displayed the corrections in horizontal range at the altitude 1,500 meters calculated above are displayed. The second row displays the corrections in range at the altitude 1,500 meters as they are shown in the mountain range tables of the 122mm Russian cannon, Mod. 1960.

Table 2: Corrections in range at the altitude of 1,500 meters; range, 12,000

Change in range for the change in initial speed of 1%	Change in range for a change in propellant temperature of 10 degree	Change in range for a change in range -wind of 10m/s	Change in range for a change in cross-wind of 10m/s	Change in range for a change in the departure angle of 0.06 degree
165m	214.3m	99m	59m	56
165m	215	100m	60m	54

4.7. Effect of the Gravitational Acceleration

The systems of differential equations (2.1.3)-(2.1.5) are obtained assuming:

- the projectile is a point-mass that does not rotate.,
- the gravitational acceleration is constant in magnitude and direction, equal to the average value $g=9.80665 m/s^2$.
- the firing site is flat.
- the Earth does not rotate.

In exterior ballistics, for relatively short firing ranges, the gravitational acceleration at the sea level is considered constant with an average magnitude of

$$g=9.80665 m/s^2.$$ (4.7.1)

As a matter of fact, due to the Earth's rotation, the gravitational acceleration depends on the (geodetic) latitude. The gravitational acceleration can be estimated by the approximate formula:

$$g_0 = 9.80665 - 0.0337 \cdot (\cos^2 L).$$ (4.7.2)

where L is the altitude of the firing site.

For firing ranges till around 50,000 meters, the influence of the change in the magnitude and direction of the gravitational acceleration can be ignored.

Thus, for ranges up to around 50,000 meters, the gravitational acceleration at the sea level is assumed to have the value given by (4.7.1).

For relatively large firing ranges (high -altitude trajectories) the gravitational acceleration along the projectile trajectory changes in magnitude according to the relation [86]:

$$g = g_0 \cos^2 \theta \cdot (1 - \frac{2y}{R}),$$ (4.7.3)

where y is the altitude of the projectile over the firing site at a moment during the flight, and R is the average radius of the Earth, $R = 6.371 \times 10^6 m$.

For customary artillery firing ranges, the influence of the change of the magnitude and the direction of the gravitational acceleration in the range of fire is insignificant and can be neglected. The gravitational acceleration is assumed to be $g=9.80665 m/s^2$.

The gravitational acceleration at the firing site over the sea level can be estimated using the following formula:

$$g = 9.80665 \cdot (1 - \frac{2y}{R}),$$ (4.7.4)

y is the altitude of the firing site.

In the following table, it is shown the gravitational acceleration as a function of the altitude of the firing site computed employing (4.7.4) is shown.

Altitude	0	1000	2000	3000	4000	5000
g	9.80665	9.80357	9.80005	9.7974	9.79434	9.7913

For relatively large artillery firing ranges, on high mountains, the change of the magnitude of the gravitational acceleration with

---

86. Whelan, P. M., Hodgson, M. J. "Esential Principles of Physics", p.155, J. Murray 1978

the altitude of the firing site should be considered in calculations, substituting g with the value given by (4.7.4) in the system of differential equations (2.1.3)-(2.1.5).

Example 7.1 Range Correction

A projectile of the Russian 122mm cannon 122mm, Mod.1960 is fired at the altitude of 3,000 meters over the sea level with a speed of 885 m/s at an angle 29.7 degrees..

The temperature, the pressure, and the water vapor pressure at the shooting site are respectively -3 degrees Celsius, 520 mmHg, and 2.88mmHg. Consider that the temperature of the propellant is 15 degrees Celsius. Assume that there wis no wind.

Find the range of the projectile considering the gravitational acceleration:

(a) $g=9.80665 m/s^2$.
(b) $g = 9.7913 m/s^2$.

Solution:

(a) Using the PC program Ran122.Bas for the given data, we find the horizontal range $x_T = 26260m$.
(b) Using the PC program Ran122.Bas, but changing the gravitational acceleration to $g = 9.7913 m/s^2$, we find $x_T = 26270m$.

Note that the above example shows that the change in range is negligible.

4.8. Effect of the Earth Curvature in Firing Range

The systems of differential equations of the projectile flight (2.1.3)-(2.1.5), and (2.3.14)-(2.3.16), are used to calculate the trajectory elements considering that the firing site is flat. Because of the curvature of the Earth, the range of fire is greater thatn the range calculated using the differential equations of the projectile flight (figure 13).

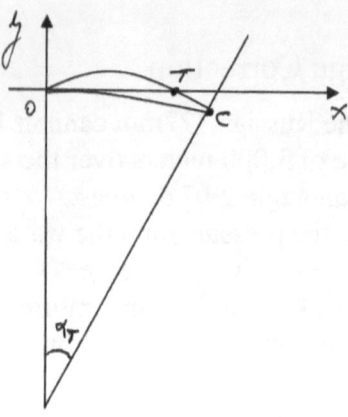

Figure 13

The real value of the projectile range represents the distance from the muzzle of the firearm to the target along the Earth's surface. The coordinates of the point of impact with respect to the coordinate system (that can be obtained from the geometry of figure 13.) are approximately:

$$x_c = x_T + \frac{x_T^2 \cot|\alpha_T|}{R}, \qquad y_c = -\frac{x_T^2}{2R}. \tag{4.8.1}$$

where x_T, α_T, and R are respectively the horizontal range, the corresponding terminal angle, and the radius of the Earth, $R = 6.371 \times 10^6 m$.

The coordinate of the point of impact can be easily introduced in the PC programs Run122.Bas, etc.

Example 8.1 Correction for the Curvature of the Earth

A projectile of the Russian 122mm cannon 122mm, Mod.1960 is fired at the sea level with a speed of 885 m/s at an angle 25.6667 degree in the TSA atmosphere.

The range table shows that the horizontal range is 20,000 meters, and the terminal angle is -46 degrees.

What is the actual range considering the curvature of the Earth?

Solution:

Substituting in (22.1.5), we find that the actual range is

$$x_c = x_T + \frac{x_T^2 \cot|\alpha_T|}{R} = 20000 + \frac{(20000)^2 \cot|-46°|}{6.731 \times 10^6} = 20057.39m.$$

The projectile will hit the target on the ground at 57.39 meters faurther than the calculated horizontal range.

We need to correct (increase) the departure angle using the approach shown in preceding sections.

4.9. Effects of the Projectile Rotation

The differential equations of the projectile trajectory are obtained assuming that the axis of symmetry of the projectile exactly overlap the tangent line to the trajectory of flight. As a result, the drag force is directed in the opposite direction of the projectile velocity.

Actually, since the moment that the projectile is launched from the firearm muzzle, the projectile rotates against the axis of symmetry of the projectile, which forms a small angle δ (attack angle) with the tangent line (velocity vector) to the trajectory. The projectile is considered a fast-spinning symmetric rigid-body.

The drag force forms an angle with the projectile axis of symmetry that usually is greater than the angle of attack. The drag force is exerted at a point that is between the projectile nose and the center of mass of the artillery projectile (or bullets). (Note that for mortar projectiles, the center of the drag force is behind the center of mass point.).

The drag force exerts an overturning moment on the projectile and tends to increase the attack angle, but because of the gyroscopic property of the spinning projectile, the projectile will rotate against an imaginary axis that tends to follow the direction of the instant velocity.

The precession motion of the axis of the projectile against a dynamic axis that is to the right of the tangent line, constrains the projectile to follow the trajectory of flight and at the same time to deviate to the right of the firing plane xy.

As a summary the spin given to the projectile at the bore of the firearm gives the projectile the stability in flight and restrict it to hit the target with its nose and, at the same time, to deviate from the launching plane to the right when the rotation of the projectile is to the right (to the left when the spin is to the left-handed).

The deviation of the projectile from the launching plane as result of the precession motion of the projectile can be estimated using the experimental formula

$$z = k \cdot t_T^2, \tag{22.1}$$

where k is an experimental constant and t_T is the time of flight to the target.

Other effects related with the spinning projectile are the Magnus effect, and the spin-damping effect.

The mathematical description of the motion of the spin-stabilized projectiles is not been treated, and it is not within the scope of the book.

References

1. Alger, R. Ph., *Exterior Ballistics*, The Lord Baltimore Press, 1906
2. Cranz, C., and Becker, K.,. *Exterior Ballistics,.* London, 1921
3. Cronander, H.A.N. S., "G-Dragfunctions.xls" and; "G-Drag-models.xls.", http://www.cronander.net/, November 10th, 2005.
4. De Mestre, N.,. *The Mathematics of Projectiles in Sport,.* Cambridge University Press, 1990.
5. DND Canada. *Field Artillery,.* Vol.ume 6,. DND Canada, 1992.— http://www.scribd.com/doc/4934783/BALLISTICS-AND-AMMUNITION,. (Web accessed December 20, 2009)
6. Dyckmans, G., *Fundamentals of Ballistics,.* (2007-2009),. e-book, https://e-ballistics.com/ebook.
7. Gubinim, S. G., and Gorovim,. S. A. *Ballistics, Handbook,* http://www.ssga.ru/AllMetodMaterial/metod_mat_for_ioot/metodichki/ballistica/index.htm.
8. Hatcher J. S., *"Hatcher's Notebook",.* Stackpole Books, 1962.
7. Hayden, R, Almgren , T., Thomas, K., and McDonald, W. T., "Lessons Learned from Ballistic Coefficient Testing,." *Exterior Ballistics Explained,.* http://www.exteriorballistics.com/ebexplained/5th/24.cfm (Web accessed July 2009.
8. Herrmann, E. E.,. *Exterior Ballistics,* . U.S. Naval Institute, The College Press, 1935.
9. Hurley, J. P., and Garrod, C., Principi Di Fisica, Zanichelli, 1986.
10. Klimi, G., Exterior Ballistics with Applications—: Skydiving, Parachute Fall, Flying Fragments,. Xlibris, 2008.
11. Klimi, G., Exterior Ballistics of Small Arms,. Xlibris, 2009.

12. Krčmář, Jan. PC Program Ballistica2.2,. http://www.balistika. cz/eng/exterior.html (accessed on 6 November 6, 2009).
13. Litz, B., "Applied Ballistics for Long Range Shooting," Applied Ballistics,. LLC, 2009.
14. McCoy, R. L., Modern Exterior Ballistics,. Schiffer Publishing Ltd., 1999.
15. McDonald, W., Inclined Fire,. June, 2003,. http://www.exterior ballistics.com/ebexplained/5th/50.cfm
16. McShane, E. J., Kelly, J. L., and Reno, F., Exterior Ballistics,. The University of Denver Press, 1953.
17. Ministry of Defense of Albania. Mountain Range Table of 122mm Cannon Mod. 1960—Projectile OF-472, Ministry of Defense of Albania, Tirana, 1972.
18. Ministry of Defense of Albania, Range Tables of Cannon 122mm, Mod. 1960,. Ministry of Defense of Albania, Tirana, 1967.
19. Mori, E., Balistica teorica e pratica;. http://www.earmi.it/ balistica , November 2009.
20. Mucinov, S. S., and Shevcenko, N. A., Zadacnik po Osnovami Strelbi is Strelkovogo Oruzie,. 1964.
21. Okunev, B. H.,. Fundamentals of Ballistics,. Vol.1, Book 2,. Moscow, 1943.
22. Ricciardelly, S., Steve's Page,. http://stevespages.com/bc.html. (web accessed July, 2009).
23. Rinker, R. A.,. Understanding Firearm Ballistics,. 6th ed. Mulberry House Publishing, 6th Ed, 2005.
24. Roller, D. E. and, Blum, R., Fisica,. Vol. 1,. Zanichelli, 1984.
25. Schaefer, J. C., A Short Course in External Ballistics,. http://www.frfrogspad.com.
26. Schaefer, J. C. *A Brief Discourse on Ballistics Coefficients*,. www.frfrogspad.com/extbal.htm (Web accessed November, 2009).
27. Shapiro, J. M., *Vneshnaja Balistika*,. Oborongiz, 1950'.
28. Whelan, P. M., and Hodgson, M. J., *Essential Principles of Physics*,. J. Murray, 1979.
29. Wikipedia,. *External Ballistics*,. http://en.wikipedia.org/wiki/ External_ballistics (accessed October 24, 2009)
30. Zill, D. G., and Cullen, M. R.,. *Differential Equations with Boundary-Value*,. 5th ed., Books/Cole, 2001.

Index

The letter *t* following a page number denotes a table, and the letter *n* denotes a footnote followed by the corresponding footnote number, if there are several notes.

www.ingramcontent.com/pod-product-compliance
Lightning Source LLC
Chambersburg PA
CBHW031815170526

45157CB00001B/69